NEURODEVELOPMENTAL DISEASES - LABORATORY AND CLINICAL RESEARCH

NEURAL TUBE DEFECTS

PREVALENCE, PATHOGENESIS AND PREVENTION

NEURODEVELOPMENTAL DISEASES - LABORATORY AND CLINICAL RESEARCH

Additional books in this series can be found on Nova's website under the Series tab.

Additional e-books in this series can be found on Nova's website under the e-book tab.

NEUROSCIENCE RESEARCH PROGRESS

Additional books in this series can be found on Nova's website under the Series tab.

Additional e-books in this series can be found on Nova's website under the e-book tab.

NEURODEVELOPMENTAL DISEASES -
LABORATORY AND CLINICAL RESEARCH

NEURAL TUBE DEFECTS

PREVALENCE, PATHOGENESIS AND PREVENTION

ALBERICH KLEIN
EDITOR

New York

Library of Congress Cataloging-in-Publication Data

ISBN: 978-1-62417-892-4

Library of Congress Control Number: 2012956343

Published by Nova Science Publishers, Inc. † New York

Contents

Preface

Neural Tube Defects (NTDs) are congenital malformations of the central nervous system caused by partial or complete failure of the neural tube to close, within 28 days of conception. They represent the second most frequent cause of congenital abnormality. In this book, the authors discuss the prevalence, pathogenesis and prevention of neural tube defects. Topics include preconception folic acid as a primary prevention tool for NTD and its risk assessment; understanding and preventing folic acid-resistant neural tube defects; knowledge of the cellular mechanisms responsible for normal neural tube formation in comprehending and preventing NTDs; developmental delay in interneuron, motor neuron, sensory tract, and synapse formation in chick spina bifida; and an examination of whether neural tube defects lead to structural alterations in the urinary and genital organs in human fetuses.

Chapter 1 – Neural Tube Defects (NTDs) are congenital malformations of Central Nervous System caused by partial or complete failure of the neural tube closure, within 28 days of conception. They represent the second most frequent cause of congenital abnormality. In most areas of USA the birth prevalence of NTDs varies from 0.8/1000 birth to 3.5/1000 birth in Mexico, whereas in European countries ranges between 0.4-2/1000 pregnancy. Anencephaly and myelomeningocele (spina bifida, SB) are the most common forms of NTDs and result from failure of the neural tube to close properly in the developing brain or in lower spine, respectively. Infants with anencephaly are stillborn or die shortly after birth, whereas children born with SB survive, but they suffer of disabilities and chronic health care issues for the rest of their life. Therefore many women choose pregnancy termination if the NTD is prenatally diagnosed. In the last two decades clinical investigations and research studies indicate that these birth defects have a multifactorial etiology and they arise from genetic and environmental interactions. Randomized controlled trials have conclusively shown that folic acid supplementation can prevent up to two-thirds of NTDs pregnancies; therefore, in order to minimize NTD risk, fertile women planning a pregnancy are recommended to assume 400μg folic acid/day at least one month before and during the first three months of pregnancy. In addition, maternal use of anticonvulsants medications or other folic acid antagonist increases the occurrence of NTD in offspring. In 1998, USA and Canada implemented mandatory folic acid fortification of wheat flour and corn and it has been reported that fortification has reduced the number of births complicated by NTD of 20% in one population-based study and 50% in two other studies. Actually, over 60 countries practice mandatory folic acid fortification of staples prevalently in America and Australia but also in Africa and in Middle East. To date, no European country has implemented mandatory folic acid fortification of staple food.

Nevertheless, the complexity of NTD genetic is underscored by the more than 150 mouse genes implicated in the pathogenesis of NTD which, for the most part, do not overlap with the folate metabolic pathway. Infact many of these genes play a role in signaling pathways acting during development, such as Planar Cell Polarity pathway that is involved in cell morphology and differentiation. The recent identification of rare pathogenic variants in human cohort of many genes belonging to the Planar Cell Polarity pathway, suggests that such genes may contribute to humans NTD pathogenesis. Societal and commercial applications of all these information are manifold, and knowledge gained in these years will yield significant benefits to the genetics and birth defects research communities, with particular implications for genetic counseling services. Furthermore, dissemination of the discoveries will likely have a substantial impact on health policy and prevention programs for NTD.

Chapter 2 – Neural tube defects (NTDs) are birth anomalies resulting from the embryonic neural tube's failure to fuse near the 28th day after conception. Several studies have demonstrated that periconception Folic Acid (FA) supplementation decreases the risk of NTDs by about 70 %. As a result, is recommended to consume 0,4 mg folic acid per day during the periconception period to reduce the risk of first occurrence NTDs. But several maternal conditions, increasing the risk of congenital NTDs, need a preconception evaluation and an appropriate intervention. It's well known that certain medications (such as antiepileptics and other anti-folate drugs) can to increase the risk of NTDs, as well as alterations in maternal folate and homocysteine metabolism. An augmented risk has been also documented in maternal diabetes and obesity, probably due to a sustained state of hyperglycemia and/or hyperinsulinemia. In addition, some studies have suggested that periconception maternal smoking can increase the risk of NTDs in offspring. Women suffering from malabsorption syndrome also represent a high-risk population, such as patient affected by NTDs or with previous pregnancies with NTDs. Some studies have demonstrated that maternal occupational exposure to organic compounds during early pregnancy can also represent an important risk factor. In conclusion, primary prevention of NTDs not only includes preconception FA supplementation but requires an adequate multidisciplinary risk assessment.

Chapter 3 – Neural tube defects (NTDs) are among the commonest birth defects, occurring in 0.5 to 2 per 1000 pregnancies. The neural tube is the embryonic structure that develops into the brain and spinal cord: the defects arise from an improper embryonic neural tube closure in the third and fourth weeks of pregnancy, causing malformations of the brain and spine (most commonly anencephaly and spina bifida).

Peri-conceptional supplementation with folic acid can prevent a proportion of NTD cases, however about 30% of NTDs are independent from folate intake. Reports of NTD recurrence in families despite high dose of folate intake, further highlights the heterogeneity of NTD etiology, suggesting the existence of different NTD subtypes, many of them being folate-sensitive, and other folate-resistant. In those 'folate-resistant' cases no preventive therapy is available.

Experimental approach in mice revealed that NTDs in Pax3 and Cited2 mutant mice should be considered folate-sensitive, while NTDs in curly tail mutant mice are folate-resistant. Inositol is a vitamin required for the neural tube to close in mice, through activation of protein kinase C (PKC). Both D-chiro-inositol and myo-inositol prevent NTD pregnancies in curly tail mice, suggesting that its use should be effective in reducing experimental folate-resistant NTD recurrence.

Moreover, inositol supplementation significantly decreased the frequency of embryonic NTDs in diabetic rats. Inositol therapy appears safe in mouse pregnancy, without fetal defects or increased fetal loss.

Subsequently, lower inositol blood concentration was found in pregnant women carrying NTD fetuses, and inositol supplementation has been associated with healthy offspring in mothers at risk for folate-resistant NTDs.

Inositol therapy in humans has been used in treating psychiatric disorders, polycystic ovary syndrome without side effects. Children have been treated with inositol for autism and respiratory distress syndrome without side effects.

High dosage of inositol was used: up to 18 g daily in adults, and 200 mg/kg in children.

Inositol is an isomer of glucose, is widespread in many foods, and is a constituent of living cells. It seems unlikely that exogenous inositol theraphy will pose a risk to the mother and/or the embryo/fetus..

Until now, 25 at high risk pregnancies have been successfully treated with myo-inositol and folic acid, without any side effects to the mother or the fetuses. More than 90 women were treated with up to 2gr inositol in the periconceptional period and reported no side effects.

All these evidences support the effectiveness and safety of periconceptional use of inositol associated with folic acid to further reduce the incidence of human NTDs.

Chapter 4 – Knowledge of the cellular mechanisms responsible for normal neural-tube formation is essential to comprehending the genesis of neural-tube defects (NTD), as well as their prevention. Neurulation is the early embryonic process in which the neural plate undergoes shaping and folding to form the neural-tube, the precursor of the central nervous system (CNS). NTD result from a failure of the rise and fusion of the neural folds, which can affect the brain and/or spinal-cord. The mechanism of NTD genesis has not yet been established, but there is consensus in the literature that it involves disruptions in neural and also mesenchymal tissues. These defects have a multifactorial etiology, a combination of genetic predisposition and/or environmental factors. In humans, supplementation with folic acid (FA) during preconception and early pregnancy prevents these defects. Dietary FA deficiency results in increased levels of homocysteine (Hcy) in the plasma. Hcy is a sulfur-containing amino acid derived from the homocysteine-methionine cycle and from the transsulfuration pathway to cysteine. In the hyperhomocysteinemia condition, the reduced availability of FA compromises nucleic-acid synthesis, which is necessary to maintain the rapid and intense cell proliferation, as well as the synthesis and expression of the genes and proteins that occur during early embryogenesis. Experimental studies have focused on the teratogenic role of Hcy in different animal models, demonstrating its effects on the occurrence of congenital anomalies. The main aim of this chapter is to describe the effects of FA and Hcy imbalance on the cellular dynamics of CNS development, using embryos of *Gallus domesticus* as the experimental model. The authors showed that Hcy induces different disruptions in external morphology, mainly NTD. Regarding the cellular organization of the CNS, Hcy-treated embryos exhibited a reduction in the thickness of the neuron-glial layers. In summary, these results demonstrate that the imbalance between FA and Hcy levels can lead to disruptions in CNS and mesenchyme development, affecting proliferation and cell adhesion, and consequently changing the arrangement of the neuron-glial layers.

Chapter 5 – Spina bifida aperta (SBA) is a congenital malformation of the spinal cord with complications such as spinal ataxia and bowel and bladder dysfunction. It is believed that a primary neurulation disorder with failed closure of the neural plate contributes to SBA. The authors developed a chick model of surgery-induced SBA that shows spinal ataxia after hatching. Incision over the length of seven somites caused SBA-like malformations in 100% of hatched chicks. This model may facilitate studies of the pathological mechanisms that lead to leg dysfunction in SBA chicks as follows. (1) Functionally, the SBA chicks exhibited leg muscle paralysis or imbalance between agonist and antagonist leg muscles. Lesions in these SBA chicks were located in the spinal segments that give rise to motor neurons innervating the dysfunctional muscles. (2) Histological analyses revealed fewer interneurons at the site of the lesion in SBA chicks than in normal chicks, but no difference was observed in the number of motor neurons. Leg dysfunction in this SBA model may be attributable to the smaller number of interneurons in the spinal segments that contain motor neurons innervating the dysfunctional muscle. (3) Immunohistochemical staining with a motor neuron marker showed that a defect in the production and elimination of excess motor neurons during the early developmental stages in SBA chicks may be involved in the pathological mechanism of the motor complications of this disease, although the final number of motor neurons was not different. (4) Histological analyses of spinal cords showed that the area of the dorsal funiculus was smaller in hatched SBA chicks than in the normal controls. Additionally, compared with normal controls, the SBA group had fewer nerve fibers in the dorsal funiculus. The ascending sensory nerve pathway was revealed by tracing the degenerated nerve fibers using osmification. In control chicks, degenerated sensory nerve fibers were observed only in the ipsilateral dorsal funiculus. In contrast, degenerated sensory nerve fibers were observed in all funiculi of SBA chicks. Consequently, fewer sensory nerve fibers ascended to the thoracic dorsal funiculus in the SBA chicks compared with normal controls. (5) Morphometric analyses of synapses and extracellular space on large montage electron micrographs of the spinal cord revealed a delay in synaptogenesis and tissue maturation in SBA chicks. The present results show a decrease in the number of interneurons and synapses and abnormal development in the motor neuron and sensory tracts of SBA chicks. These findings suggest that the complex neural networks controlling the muscles developed until the late gestational period and indicate the necessity for early therapy to prevent SBA-induced neurological deficits.

Chapter 6 – Purpose: Anencephaly is the most severe neural tube defect in human fetuses. The objective of this paper is to analyze the structure of the ureter, bladder and penis in anencephalic human fetuses.

Methods: The authors studied 40 normal human fetuses (20 male and 20 female, aged 14 to 23 WPC) and 12 anencephalic fetuses (5 male and 7 female, aged 18 to 22 WPC). The ureter, bladder and penis were removed and processed by routine histological techniques. Stereological analysis of collagen, elastic system fibers and smooth muscle was performed in sections. Data were expressed as volumetric density (Vv-%). The images were captured with Olympus BX51 microscopy and Olympus DP70 camera. The stereological analysis was done using the software Image Pro and Image J. For biochemical analysis, samples were fixed in acetone, and collagen concentrations were expressed as micrograms of hydroxyproline per mg of dry tissue. Means were statistically compared using the unpaired t-test ($p < 0.05$).

Results: The authors observed a significant increase ($p<0.0001$) in the Vv of collagen in the bladders of anencephalic fetuses (69.71%) when compared to normal fetuses (52.74%), and a significant decrease ($p<0.0001$) in the Vv of smooth muscle cells in the bladders of anencephalic fetuses (23.96%) when compared to normal fetuses (38.35%). The biochemical analyses showed a higher concentration of total collagen in the bladders of anencephalic fetuses (37354μg/mg) when compared to normal fetuses (48117μg/mg, $p<0.02$). Intracavernosal septum was present in all samples. The authors did not observed differences in the Ad of Cot and SMC in the penises of anencephalic fetuses when compared to normal ones. Performing simple Linear Regression, it can be suggested that during the human development, there is a gradual increase in Cot ($R^2= +0.45$) and a decrease in SMC ($R^2=-0.62$) in both groups studied. Elastin was observed only in fetuses from 20th WPC. There was no difference between gender, anencephalic and control groups, when calculating age, weight, ureteral wall thickness and lumen area. Using simple linear regression we could observe that ureteral lumen area ($R^2 = 0.52$) and wall thickness ($R^2=0.73$) increases gradually with ages advance, independent of the presence of anencephaly. This relation was not observed when using weight as a parameter ($R^2= 0.31$).

Conclusions: The structural alterations of the bladder found in this study may suggest the existence of functional alterations in the bladder of anencephalic human fetuses. There was no difference in the structure of the genitalia of anencephalic fetuses compared to normal ones. Elastin is documented from 20th WPC wich suggests the maintenance of erectile function. Histochemistry and immunolabeling suggests that penile shaft development is maintained and unaltered in NTD fetuses. Although anencephaly is a disease that is incompatible with babies' long spam survivors, this present study shows that ureteral development is unchanged. In our opinion, its anatomy did not suffer any alteration in consequence of this disease and, in the future, this tissue could be used as a substitute for native tissue.

In: Neural Tube Defects
Editor: Alberich Klein

ISBN: 978-1-62417-892-4

Chapter I

Prevalence, Pathogenesis and Prevention of Neural Tube Defects

Patrizia De Marco, Elisa Merello and Valeria Capra [*]
Neurosurgery Department, G. Gaslini Institute, Genova-Italy

Abstract

Neural Tube Defects (NTDs) are congenital malformations of Central Nervous System caused by partial or complete failure of the neural tube closure, within 28 days of conception. They represent the second most frequent cause of congenital abnormality. In most areas of USA the birth prevalence of NTDs varies from 0.8/1000 birth to 3.5/1000 birth in Mexico, whereas in European countries ranges between 0.4-2/1000 pregnancy. Anencephaly and myelomeningocele (spina bifida, SB) are the most common forms of NTDs and result from failure of the neural tube to close properly in the developing brain or in lower spine, respectively. Infants with anencephaly are stillborn or die shortly after birth, whereas children born with SB survive, but they suffer of disabilities and chronic health care issues for the rest of their life. Therefore many women choose pregnancy termination if the NTD is prenatally diagnosed. In the last two decades clinical investigations and research studies indicate that these birth defects have a multifactorial

[*] E-mail: valeriacapra@ospedale-gaslini.ge.it.

etiology and they arise from genetic and environmental interactions. Randomized controlled trials have conclusively shown that folic acid supplementation can prevent up to two-thirds of NTDs pregnancies; therefore, in order to minimize NTD risk, fertile women planning a pregnancy are recommended to assume 400µg folic acid/day at least one month before and during the first three months of pregnancy. In addition, maternal use of anticonvulsants medications or other folic acid antagonist increases the occurrence of NTD in offspring. In 1998, USA and Canada implemented mandatory folic acid fortification of wheat flour and corn and it has been reported that fortification has reduced the number of births complicated by NTD of 20% in one population-based study and 50% in two other studies. Actually, over 60 countries practice mandatory folic acid fortification of staples prevalently in America and Australia but also in Africa and in Middle East. To date, no European country has implemented mandatory folic acid fortification of staple food.

Nevertheless, the complexity of NTD genetic is underscored by the more than 150 mouse genes implicated in the pathogenesis of NTD which, for the most part, do not overlap with the folate metabolic pathway. Infact many of these genes play a role in signaling pathways acting during development, such as Planar Cell Polarity pathway that is involved in cell morphology and differentiation. The recent identification of rare pathogenic variants in human cohort of many genes belonging to the Planar Cell Polarity pathway, suggests that such genes may contribute to humans NTD pathogenesis. Societal and commercial applications of all these information are manifold, and knowledge gained in these years will yield significant benefits to the genetics and birth defects research communities, with particular implications for genetic counseling services. Furthermore, dissemination of the discoveries will likely have a substantial impact on health policy and prevention programs for NTD.

1. Prevalence of Neural Tube Defects (NTDs)

Neural Tube Defects (NTDs) are a group of congenital anomalies that occur when the normal neural tube closure process fails around the 28th day after conception (Sadler, 1998). The most common classification schema for NTDs includes anencephaly, SB, and encephalocele. Craniorachischisis and iniencephaly are considered to be rarer disorders in most populations. Anencephaly involves absence of the skull, with the cerebral hemispheres completely missing or reduced to small masses attached to the base of the skull. SB (which includes myelomeningocele and meningomyelocele) is defective closure of the bony encasement of the spinal cord, through which the cord and meninges may or not protrude. Encephalocele is the protrusion of some or all of the brain through a defect in the skull. NTDs group also includes other conditions, broadly described as covered-skin (occult) spinal dysraphisms (lipoma, tight filum terminale) and some complex dysraphic states as diastematomelia, spinal segmental dysgenesis, and caudal regression syndrome (Carter et al. 1976; Rossi et al. 2004).

Prevalence is defined as the number of affected persons in the population at a specific time divided by the number of persons in the population. NTDs have been described since ancient times around the world (Elwood et al. 1992). Ancient skeletons with NTDs were found in North America and in Africa. Egyptian mummies also present with SB. NTDs are dramatic birth defects, whose prevalence is impossible to establish since 1970, when prevalence estimation become a relevant starting point of national public health services for adoption of prevention strategies. NTDs are listed as a "rare diseases" by the Office of Rare Diseases (ORD) of the National Institutes of Health (NIH) (http://rarediseases.info.nih.gov/RareDiseaseList.aspx). Orphanet, the reference portal for information on rare diseases and orphan drugs led by a European consortium of around 40 countries, also defines NTDs as a rare condition (http://www.orpha.net/consor/cgi-bin/index.php).

Case ascertainment is crucial for the correct estimation of NTDs prevalence. In fact, NTDs are birth defects with a high rate of mortality (100% of anencephaly cases and the half of the cases of SB) (Nakano, 1973; Bowman et al. 2001). In addition, fetuses with NTDs are probably aborted spontaneously before the pregnancy is recognized. In Western and developing countries, prenatal diagnosis combined with induced abortion is used to reduce the incidence of these devastating defects. It is therefore difficult to deduce natural prevalences, since that the number of electively aborted cases or spontaneously aborted is unknown. Studies showed that elimination of cases that are aborted would estimate very low prevalences (Ethen and Canfield, 2002). The total prevalence is equal to the birth prevalence only in populations where elective termination of pregnancy is not permitted.

Coding and classification methods are another critical issue for the epidemiology of NTDs. Among the most comprehensive classifications of congenital abnormalities, is that contained in the ICD, the International Classification of Diseases. It is arranged by system, part, and organ, and includes virtually every deviation from normal, originating prenatally, present at birth, or attributable to conditions present at birth, regardless of medical importance or aetiological status. Based on ICD, the category of "Neural Tube

Defects" includes three conditions with code 740 (anencephalus and similar anomalies), 741 (SB, excluding spina bifida occulta) and 742 (encephalocele) (International Classification of Diseases, 1998). The ICD classification code does not differentiate on the basis of the site of the lesions.

There are additional concerns regarding the classification of NTDs, such as the occurrence of NTDs in isolation or with birth defects affecting other organ systems. The best choice would be to separate cases into "isolated" and "multiple" or "syndromic" cases to gain insights into the roles of risk factors and environmental exposures (Khoury et al. 1992; Rasmussen et al. 2003).

Birth prevalence can be estimated from a number of sources, such as birth registries and hospital records, but the best estimates are those that include on both live births and stillbirths, although these data are difficult to retrieve.

If the data are provided from a hospital they will include only live birth or possibly fetal death events. Prevalence studies and registries provide data around the world and they may have different purposes. The most common purpose is epidemiologic surveillance of defects, particularly for detection of new potent teratogens. For EUROCAT, a European network of population-based registries for the epidemiologic surveillance of congenital anomalies, the prevalence is shown for live births plus fetal deaths plus induced abortions (EUROCAT Working Group, 1991; 1997; 2002). Otherwise for the International Clearinghouse for Birth Defects Surveillance, a birth defect surveillance and research programmes from around the world, the prevalence is provided for live births and stillbirths (International Clearinghouse for Birth Defects Monitoring Systems, 2002).

Over the last several decades, women carrying a fetus with an open NTDs have been found to have elevated maternal serum levels of alpha-fetoprotein and reduced levels of human chorionic gonadotropin and unconjugated estriol (Canick and Saller, 1993). Prenatal screening of these substances, along with ultrasonography (Vintzileos et al., 1987), have allowed NTDs to be identified *in utero*. Epidemiological data from the period before *in utero* diagnosis and termination of affected pregnancies established two trends. The first illustrated epidemics of the incidence of NTDs that cycled over years, and the second established that there has been a general decline in the birth incidence of NTDs over several decades, which has been noted worldwide (Elwood and Elwood, 1980). The best example of epidemics of the occurrence of NTDs was reported from Birmingham, England. The prevalence rose from 2.0 per 1000 births in 1936–7, to 2.8 in 1940, fell to 1.54 in 1948 and peaked again in 1956 at 2.8 per 1000 (Leck, 1996). Other "epidemics" have been described in Boston, Providence and New York in the USA, Canada, Berlin and in Western

Germany, and in a number of other areas (Elwood and Elwood, 1980). Overall, rates for anencephaly and spina bifida have declined in North America, Europe, Australia, and New Zealand (McDonnell et al., 1999; Center for Disease Control and Prevention, 1992; Yen et al., 1992; Campbell et al., 1986; Little and Elwood, 1992). The proportion of NTDs that were electively terminated increased with time in some regions studied (Whiteman et al., 2000; Bower et al., 1993; EUROCAT Working Group, 1991). Thus, some of this decline in NTDs prevalence may be attributable to increased prenatal diagnosis and elective termination of NTDs-affected pregnancies. However, this trend was observed prior to the widespread use of prenatal diagnosis and has occurred in areas where elective termination is not allowed.

Folic acid (FA) taken orally on a daily basis is shown to lower the occurrence and recurrence of NTDs in their own offspring and in their relatives. The Medical Research Council was the first to prove conclusively that when women who had a previous child affected by an NTDs took 4.0 mg of FA daily, beginning three months prior to conception, there was a 70% reduction of the recurrence risk (MRC Vitamin Study Research Group, 1991). These findings led to statements from the National Health about prevention of these defects by improving their dietary intake of FA and by periconceptional FA supplementation (Centers for Disease Control, 1992). Subsequently, in 1994, the NHMRC recommended voluntary fortification of various foods with folate (National Health and Medical Research Council, 1994). International data for the period from 1974 to 1988 showed marked variations among countries in the reported incidence of NTDs (International Clearinghouse for Birth Defects Monitoring Systems, 1993). In the period 1991-2000, the average annual incidence of births of anencephalus varied between 16 per 10,000 births in Mexico and 0.3 per 10,000 in Finland (International Clearinghouse for Birth Defects Monitoring Systems, 1993; 1994; 1995; 1996; 1997; 1998; 1999; 2000; 2001; 2002). The highest reported rate of spina bifida was 22.3 per 10,000 births in Palestinia and the lowest rate was 1.8 per 10,000 in Hungary (Dudin, 1997; International Clearinghouse for Birth Defects Monitoring Systems, 2002). Encephalocele varied between 3.2 per 10,000 births in Mexico and 0.3 per 10,000 in Israel. In general, high prevalences are registered from high risk areas like Cina, India, Mexico, South America, and Turkey. Prevalences are low in Africa, where the information is limited, and in Oceania (International Clearinghouse for Birth Defects Monitoring Systems, 1993; 1994; 1995; 1996; 1997; 1998; 1999; 2000; 2001; 2002). There are racial or ethnic differences in risk for NTDs. In the United States, NTDs prevalence has been reported to be highest among Hispanics, followed in descending order by non-Hispanic whites, Native Americans, African-Americans, and Asians (Owen et al., 2000; Hendricks et al., 1999; Canfield et al., 1996; Shaw et al., 1994). Differences in racial/ethnic rates may be due to differences in genetic susceptibility to NTDs, cultural behaviors, diet, or other factors. The last report of EUROCAT on trends in the total prevalence of NTDs from 1980 up to 2002, in the context of a survey in 18 European countries of periconceptional FA supplementation (PFAS) policies and their implementation showed that in the UK and Ireland, countries with PFAS policy, there was a 30% decline in NTDs total prevalence, but it was difficult to distinguish this from the pre-existing strong decline. In other European countries with PFAS policy, there was virtually no decline in NTDs total prevalence whether a policy was in place by 1999 or not. The potential for preventing NTDs by periconceptional FA supplementation is still far from being fulfilled in Europe. Only a public health policy including FA fortification of staple foods is likely to result in large-scale prevention of NTDs (EUROCAT, 2002). Overall, the fact that introduction of FA prevention policies coincides with the change in the epidemiology of NTDs caused by new methods of early diagnosis and termination of affected pregnancies makes difficult to evaluate the effect of prevention strategies. Moreover, because the maternal serum alpha-feto protein (AFP) screening test is not likely to detect skin-covered lesions, these cases could have been missed by screening and underreported to the NTDs registries , thus resulting in a lower SB prevalence.

2. Prevention of NTDs

In 1964, Brian Hibbard was the first who addressed the importance of folic acid (FA) during pregnancy, suggesting that the detection of megaloblastic anemia in mid or late pregnancy implied an antecedent defect in folate metabolism. This was supported by Smithells and others (1976) who showed that patients with megaloblastic anaemia had a high incidence of NTDs-affected pregnancies. Throughout the 1980s, a number of observational and intervention studies gave strong support to the hypothesis that FA, either alone or in multivitamin preparations, could prevent a high proportion of NTDs if taken before and during early pregnancy. Smithells and coworkers (1980a) measured the first trimester blood vitamin levels in women who had a baby with NTDs, and observed that red blood cell folate and leukocyte ascorbic acid levels were significantly lower when compared to mothers without NTDs-affected offspring. This study resulted in the hypothesis that periconceptional intake of vitamins might protect against NTDs offspring. Laurence and coworkers (1981) studied the effect of a 4 mg FA supplementation before and during early pregnancy on the recurrence of NTDs in a small-randomized trial. They showed a drop in NTDs recurrences in folate users compared to nonusers. However, the study of Laurence has been criticized, mostly because of the small numbers. The effect of multivitamin supplementation before and during early pregnancy on recurrence of NTDs was studied by Smithells and colleagues (1980b; 1983) in women with a history of one or more NTDs-affected pregnancies. The multivitamin supplement contained folate, riboflavin, ascorbic acid, and vitamin A. The obtained results were striking, and they were the first who reported primary prevention of NTDs by periconceptional intake of a multivitamin preparation. This study has been criticized because of a selection bias; the case and control groups were not entirely comparable. Several groups repeated these studies and showed a likewise effect of multiple vitamins or folate on the prevention for the recurrence of NTDs.

The strongest support for the hypothesis of the role of FA in NTDs-affected pregnancies came from two large, randomised studies. The first one was a multicenter study performed by the British Medical Research Council, examining the effect of periconceptional folate and multivitamin supplements in about 1200 high-risk mothers who have already had a NTDs-affected child (MRC Vitamin Study Research Group, 1991). These women were randomly assigned to four groups allocated to receive the following regimens of supplementation. Group one received daily 4 mg of periconceptional FA; group two received the same, but additionally a multivitamin supplementation; group three received the vitamin supplementation without FA; and the final group received a placebo containing only minerals. This study showed that FA alone was as good at preventing recurrences of NTDs as the multivitamin plus FA. The recurrence of NTDs pregnancy was reduced by 72% if compared with placebo-treated women. Vitamins without FA and minerals alone were ineffective in preventing NTDs. The largest and most important randomized-controlled study of prevention on the first occurrence of NTDs by vitamins was performed by Czeizel and Dudas in Hungary (1992). This study lasted eight years and involved more than 4000 women. These women were subdivided in two groups: one received a multivitamin containing 0.8 mg of FA, 11 other vitamins, and trace elements, and the other group received a placebo multimineral containing only the trace elements. The results showed six children with NTDs in the latter group and none in the first group. This study demonstrated a complete prevention of the first occurrence risk of NTDs due to periconceptional FA.

In a more recent and wide campaign, Berry and collaborators in 1999 carried out a trial in a high-risk area, like China, among 130,142 women who took FA at any time before or during pregnancy and 117,689 women who did not take FA. They demonstrated a reduction of 79% of NTDs-affected pregnancies using an amount of 400 μg/day of FA.

Since 1991 public campaigns of prevention started in different countries worldwide and each country decided using two different approaches: public information campaigns or a mandatory fortification policy. As a result, the US Public Health Service recommended in 1992 that all women of child-bearing age consume at least 400 μg of synthetic FA daily in addition to eating a folate-rich diet. In 1996, the US Food and Drug Administration (FDA) considered FA enrichment of all cereal and grain products in the United States. This generated substantial debate as potential risks of such a broad program. A major concern was the impact on elderly patients, particularly those with vitamin B_{12} deficiency. Ultimately, the FDA mandated fortification by January 1998 at a level of 140 μg/100 g of flour. This is a level of fortification below what is needed to maximize prevention of NTDs in the majority of reproductive aged women.

Some European countries did not adopt such protocols and there are still public debate on such type of prevention. In Italy, the relatively low prevalence of NTDs, together with some underestimation of the social weight of birth defects by the National Health Service, may have contributed to the delay in establishing approaches for the prevention of NTDs. In 2004, the Italian Network for Folic Acid Promotion, organized and coordinated by the National Center for Rare Diseases of the Italian National Institute of Health (ISS), developed and diffused a recommendation promoting a voluntary increased intake of folic acid (0.4 mg/day) by women in fertile age (http://www.cnmr.iss.it). The Italian Ministry of Health in the Drugs Bulletin published and distributed the draft recommendation to family practitioners and specialists.

Overall, pre and post fortification studies carried out during the last two decades in different populations demonstrated that an increased intake of FA led to a decrease of NTDs risk; the observed decrease is not fixed for all populations but is related to the risk of the population not exposed to FA supplements; the decrease of NTDs risk does not fall below 0.6 per 1,000 births (Mastroiacovo and Leoncini, 2011).

Thus, to decrease NTDs risk, it is recommended that all women capable of becoming pregnant should have more FA. Optimal folate levels can be achieved through diet and life style, voluntary intake of fortified food, or use of FA supplements. Women living in countries where there is not mandatory fortification can choose an adequate diet or to use a supplement with FA. The most important recommendation is to monitor folate levels, in particular the concentration of red blood cell folate that must to be as more than 906 nmol/L (Pietrzik et al. 2007).

A more recent approach on prevention of NTDs is the use of inositol (Cavalli et al., 2002). In fact, the recurrence of NTDs despite supplementation of high dose of FA further suggests that a proportion of NTDs cases might be resistant to FA. Moreover, animal models suggest the high variability on the sensitivity to FA, indicating that only some sub-types of NTDs could be responsible to this vitamin. Inositol (particularly myo- and chiro-inositol) could prevent folate-resistant NTDs in the *curly-tail* mutant mouse, suggesting that some cases of human NTDs might benefit from inositol supplementation. In humans, lower inositol blood concentration was found in pregnant women carrying NTDs fetuses, whereas a periconceptional combination therapy with FA associated with inositol has been linked to normal live births, despite high NTDs recurrence risk. Inositol is a constituent of living cells and is widespread in many foods. On average each day we consume about 1g of inositol. Inositol therapy appears to be completely safe. Inositol-treated mice revealed no major fetal defects and no increase in fetal loss. Trials in adults with psychiatric disorders, in autistic children, in women with polycystic ovary syndrome and in infants with respiratory distress syndrome (Benjamin et al., 1995; Colodny and Hoffman, 1998; Levine et al., 1997; Nestler et al., 1999; Palatnik et al., 2001; Howlett and Ohlsson, 2003) have all demonstrated benefits with no side effects when the dose of inositol is less than 6 g/day. Mothers who took inositol in a third pregnancy reported no side effects. Thus, it seems highly unlikely that exogenous inositol therapy will pose a risk to the mother and/or embryo/fetus. A pilot study was carried-out on fifteen pregnancies from 12 Caucasian women with at least one previous affected pregnancy who underwent periconceptional combined myo-inositol and FA (Cavalli et al. 2011). Three women received 500 mg myo-inositol and 5 mg FA daily (Cavalli et al., 2008), while nine women received 1000 mg myo-inositol and 5 mg FA daily. Supplementation was started at least 2 months before conception, continued for at least 60 days of pregnancy, and was usually discontinued after 60 days. Maternal serum alpha-fetoprotein levels were found in the normal range, and normal results on ultrasound examination were found in all the pregnancies. No collateral effects or uterine contractions were demonstrated in this pilot study after inositol supplementation, and seventeen babies were born without any type of NTDs. In the same time, Copp et al. in UK proposed a randomised, double blind clinical trial to compare FA plus placebo, or FA plus myo-inositol (Prevention Of Neural Tube defects by Inositol in Conjunction With Folic Acid (PONTI); http://clinicaltrials.gov/show/NCT00452829), for prevention of NTDs recurrence. This will evaluate the effectiveness of inositol in the context of co-administration with FA. When such study will be completed, it will be possible to assess the safety and the efficacy of inositol in the prevention of NTDs.

3. Pathogenesis of NTDs

Neurulation, the formation of the neural tube from the undifferentiated ectoderm of the gastrulation stage embryo involves the orchestration of a series of complex morphological events. Neurulation in mammalian embryos occurs in two phases: primary and secondary neurulation (Copp and Greene, 2003; Greene and Copp, 2009). These two phases occur in distinct areas along the rostro-caudal axis of the embryo. Secondary neurulation is limited to the tail bud, which lies beyond the caudal neuropore. Primary neurulation can be subdivided into four broad morphological steps: (1) induction of the neural plate; (2) folding of the neural plate; (3) longitudinal bending of the neural plate; and (4) fusion of the neural folds (Smith and Schoenwolf 1997).

Primary neurulation generates the entire neural tube anteriorly to the caudal neuropore. During this process, occurring during the third and fourth weeks of human development (Carnegie stages or CS 8-13) the flat layer of ectodermal cells overlying the notochord is transformed into a hollow tube. Eighteen days after fertilization (CS 8), the midline dorsal ectoderm of the embryo thickens and forms the neural plate while cell shape changes. The neural plate first appears at the cranial end of the embryo and differentiates in the caudal direction. The edges of the plate thicken and begin to move upward forming the neural fold. The neural plate becomes narrower, longer, and is transformed from an elliptical to a key-hole shaped structure. This transformation occurs by polarized cell movements in the medial direction and cell intercalation in the midline. The mechanism of these movements, known as convergent extension (CE), is not specific only to the neural tube formation. On day 19 (CS 8.5), the border of the neural plate becomes gradually more pronounced and elevated. The neural plate folds longitudinally along the midline of the plate from the head toward the tail to form the neural groove. The folds rise up dorsally, approach each other and ultimately merge together, forming a tube open at both ends by day 23 (CS 10.5). As the neural folds fuse, the cells adjacent to the neural plate also fuse across the midline to become the overlying epidermis. The rostral and caudal openings are called neuropores and are best distinguished around day 23 when about 17−19 somites are visible and close, on the 26th (CS 12) and 28th (CS 13) days of gestation respectively (O'Rahilly and Müller, 2002). The number of initiation sites of fusion and their location is always been object of debate. In mouse, 3 sites of closure have been described (Fleming and Copp, 2000; Green and Copp, 2009), but the exact number of sites remains controversial, varying between 2, 3, and 5 sites (van Allen et al., 1993; Nakatsu et al., 2000; O'Rahilly and Muller, 2002). Only closures 1 and 3 were proven to definitely occur. In humans the fusion of the neural folds has originally been described as a process initiated at a single site, and extending bi-directionally, rostrally and caudally, from this initiation site to the rostral and caudal neuropores (Purves and Lichtman, 1985), even if an hypothesis of "multiple site of neural tube fusion" has been formulated also in human embryos by van Allen et al. (1993) who proposed the existence of five neural tube closure sites. According to this model, closure 1 occurs in the cervical region and extend bidirectionally. Closure 2 take place at the junction of prosencephalon and mesencephalon and also extend bidirectionally. Closure 3 initiates at the rostral tip of the neural groove and proceed caudally to meet closure 2. Closure 4 covers the rhomboencephalon and completes the closure of cranial neural tube. Finally closure 5 initiates from the caudal end of the neural groove and spreads cranially.

Open NTDs (including open SB and anencephaly in mammals) result from failure of the neural folds to appose and fuse during primary neurulation. Open NTDs can occur at different levels of the body axis, both in humans and in mutant mouse strains, and this variation in axial level reflects the occurrence of multiple closure sites (Kibar et al., 2007a; Green and Copp, 2009). Failure of closure 1 leads to craniorachischisis where the neural tube remains open throughout the brain and spinal cord. Failure of the caudal spread of fusion from closure 1 results in open SB or myelomeningocele. Failure of closure 2 leads to exencephaly that by late gestation converts into anencephaly, where the skull vault is missing and the brain tissue is degenerated. Failure of closure 3 leads to anencephaly confined to the forebrain region and that is often associated with split-face malformation. An incomplete membrane fusion of closure 4 results in occipital cephalocele and finally failure of closure 5 results in lumbo-sacral myelomeningocele (De Marco et al., 2011).

In contrast to primary neurulation, secondary neurulation occurs by proliferation of stem cells (Schoenwolf, 1984; Copp and Brook, 1989; Catala et al., 1996), which form a rod-like condensation that subsequently cavitates. The cavitation transforms the rod into a tube, and the lumen of this tube comes into continuity with the lumen of the tube formed during primary neurulation (Schoenwolf, 1984). Failure of secondary neurulation leads to some forms of closed NTDs where the developing neural tube fails to separate from other tissues of the tail bud. Dysfunctional secondary neurulation can lead to an inappropriately low-lying spinal cord. This commonly leads to tethering of the spinal cord, which may cause neuronal damage at the site of the defect, due to differential rates of elongation of the spinal cord and vertebral column as the body axis lengthens during development. Patients with such distal spinal cord anomalies, commonly classified as tethered cord syndrome (Bassuk et al., 2005) may manifest symptoms, including lower limb abnormalities and bowel and bladder dysfunction, which result from inappropriate innervation from the malformed spinal cord. It is important to note that other forms of "closed" NTDs can be caused by defects in the developing axial mesoderm at any level of the neuraxis, despite the completion of primary and secondary neurulation. Despite years of intensive epidemiological, clinical and experimental investigations, a common mechanism explaining NTDs etiology has not been defined. Epidemiological studies implicates a multifactorial etiology in which interactions between genetic and environmental factors contribute to the probability of the developing the trait. A number of maternal factors have been found implicated in the etiology of NTDs, particularly hyperthermia, diabetes, hyperinsulinemia, obesity, psychosocial, or emotional stress (Moretti, et al., 2005; Loeken, et al., 2005; Shaw et al., 1996; Shaw et al., 2003; Wasserman et al., 1998; Ray et al., 2004; Dietl 2005). Maternal epilepsy and medications used to treat epilepsy such as valproic acid and carbamazepine, have been also found to increase the risk of spina bifida (Matalon, et al 2002; Myrianthopoulos and Melnick; 1987). Other drugs (thalidomide, folate antagonists, androgenic hormones), substance abuse (alcohol), chemical agents (organic mercury, lead), maternal infections (rubella, cytomegalovirus, Toxoplasma gondii, syphilis) are capable of causing congenital malformations of the Central Nervous System structures (Shenefelt 1972; Wilson 1973; Committee on Developmental Toxicology, National Research Council 2000; Shepard et al. 2002). In addition, parental socioeconomic class, occupation (nursing, processing food and beverages, farming, textile dye and leather industries, spraying pesticides) and exposure to noxious agents such as organic solvents, anesthetic agents, paints, X-radiation, have been reported to be associated with a high risk for NTDs (Brender & Suarez 1990; Matte et al. 1993; Sever 1995; Shaw et al. 2002; Blanco Munoz et al. 2005). Other factors are still mostly speculative and need to be further evaluated (Correa, 2000; Sever, 1995; Whiteman, et al.; 2000). A positive association was also observed between SB and caffeine consumption (Schmidt, et al., 2009). However, many of the reported association have been weak and have not been replicated. More recently, the association of maternal high caffeine intake, low calorie diet, occasional consumption of fruit and vegetables, alcohol consumption and, above all, lack of folate supplementation at any time may be causative for the burden of SB in the Italian population (De Marco et al. 2011).

Several lines of evidence suggest a genetic component in the onset of the NTDs. Firstly, NTDs are associated with known genetic syndromes, trisomies 13 and 18 and various chromosome rearrangements (O'Reilly and Shields, 2000; Online Mendelian Inheritance in Man 2000); secondly, in not-syndromic NTDs, estimated recurrence risk in siblings is 2-5%, representing up a fifty-fold increased risk over that observed in general population; finally, the presence of family history in a number of affected individuals (Risch, 1990; Partington and McLone, 1995). About 70% of all NTDs are isolated and not syndromic and show multifactorial inheritance. Identification of genes predisposing to NTDs in human beings using positional cloning is difficult because of the underlying complex aetiology and the paucity of large families with multiple NTDs affected members. A genome wide linkage screen in NTDs has identified few chromosomal regions of interest; however, results from such studies were not further pursued and should be interpreted cautiously as they are complicated by probable genetic heterogeneity and variable penetrance (Rampersaud et al., 2005; Stamm et al., 2006).

Given the protective role of folate, proteins that mediate, or are functionally associated with folate metabolism have provided candidates for genetic analysis in human NTDs. There are over 25 proteins involved in the folate and closed-related homocysteine (Hcy) pathways. Originally identified as a risk factor for vascular diseases (Frosst et al., 1995), the 677C→T thermolabile isoform of *MTHFR* first emerged as a possible genetic risk factor for NTDs in some populations (van der Put, et al., 1995; Ou et al., 1996; Christensen et al. 1999). MTHFR reduces 5,10 methylenetetrahydrofolate to 5-methyl-tetrahydrofolate (the predominant circulatory form of folate) for use as a carbon donor for the remethylation of homocysteine to methionine. Low MTHFR activity is associated with low plasma folate levels and high Hcy levels (Frosst et al. 1995). Penetrance of the 677C→T is suggested to be modified by dietary and supplemental folate (Ou et al 1996). Other studies failed to demonstrate an association between *MTHFR* polymorphisms and NTDs in different populations (Koch et al. 1998; Gonzales-Herrera et al., 2002; Revilla et al. 2003). A recent meta-analysis found a pooled odds ratio for cases homozygous at 677C→T (677TT) of 1.7 (95%CI 1.4-2.2), with a pooled attributable fraction of NTDs cases of 6% for homozygosity (Botto and Yang Q, 2000). A second mutation in the *MTHFR* gene (1298A→C) has also been positively associated with NTDs (van der Put et al. 1998; Weisberg et al. 1998; De Marco et al. 2002; O'Leary et al. 2005). The 1298 variant is in strong linkage disequilibrium with the 677 variant, such that 1298C and 677T changes are rarely seen on the same allele (Stegmann et al. 1999). Therefore, the two mutations can occur on separate alleles in the same individual who may be predisposed to high Hcy levels, if folate status is low. Compound heterozygosity for the 677T and 1298C alleles (677C/T/1298A/C) has been suggested to increase the risk, although the results have been conflicting (van der Put et al. 1998). Given its important role in folate metabolism, the trifunctional enzyme methylenetetrahydrofolate dehydrogenase /methenyl tetrahydrofolate cyclohydrolase/formyltetrahydrofolate synthetase (*MTHFD1*) has been analysed in NTD pathogenesis. Brody et al. (2002) demonstrated that the polymorphism 1958G→A (R653Q) within the *MTHFD1* gene resulting in the substitution of a conserved arginine by glutamine is a maternal risk factor for NTDs. Mothers who possess two copies of the Q allele have an increased risk of an NTD-affected pregnancy. There was also a suggestion that the Q allele decreases fetal viability. The impact of the *MTHFD1* 1958G>A polymorphism on NTD risk was elevaluated in the Italian population both by case-control and family-based studies (De Marco et al. 2006). An increased risk was found for the heterozygous 1958G/A (OR=1.69; P=0.04) and homozygous 1958A/A (OR=1.91; P=0.02) genotypes in the children. The risk of an NTD-affected pregnancy of the mothers was increased 1.67-fold (P=0.04) only when a dominant effect (1958G/A or 1958A/A *vs* 1958G/G) of the 1958A allele was analysed. Family-based tests also confirmed a significant excess of transmission of the 1958A allele to affected individuals, demonstrating that this variant is a genetic risk factor for Italian NTD cases. A potential role of the mitochondrial paralogue *MTHFD1L* as candidate gene for NTDs association has been also investigated. In particular, a common triallelic deletion/insertion variant 781-6823ATT(7-9) which influences splicing efficiency was tested and a significant incresead risk for allele 1 [ATT(7)] was identified (Parle-McDermott et al. 2009). The methionine synthase (*MTR*), a vitamin B_{12}-dependent enzyme, utilizes 5-methyltetrahydrofolate, generated by *MTHF*R, for homocysteine remethylation to methionine. A polymorphism of the *MTR* gene, the 2756A→G (D919G) that lead to the substitution of an aspartic acid with a glycine residue, has been considered in several studies, but the majority have not reported a significant association (Morrison et al. 1998; Gueant-Rodriguez et al. 2003). A common variant in the gene methionine synthase reductases (*MTRR*), the 66A→G (I22M), that converts a isoleucine into a methionine residue, has been shown to confer a four-fold increased risk for NTDs in mothers having low vitamin B_{12} levels (Wilson et al. 1999). Other studies have also suggested an effect on NTDs risk, both for mothers and children with 66GG mutant genotype, but interactions with nutrients has not been examined. The *MTRR* mutant genotype combined with *MTHFR* mutant genotype conferred a significant four-fold increased risk in children (Wilson et al. 1999). An interaction of *MTR* and *MTRR* genotypes has also recently reported. Nevertheless, transgenic mice with partial or complete knock-out for the *MTR* or *MTRR* genes display no developmental NTDs phenotype, adding inconclusive evidence for the role of these genes in the pathogenesis of NTDs (Swanson 2001; Elmore 2007).

Additional studies are needed to better define the nature and extent of the interaction between folate-related genetic variants and nutrient factors in modulating the incidence and penetrance of the NTDs phenotype. A recent study of re-sequencing in 31 folate-related genes in a 480 case-control population identified two complex genetic risk signatures. Interestingly, these signatures differed by race/ethnicity: a Hispanic risk profile pointed to alterations in purine biosynthesis, whereas that in non-Hispanic whites implicated homocysteine metabolism (Marini et al. 2011). In this regard it will be imperative to replicate these findings in additional populations; particularly in non-Hispanic whites.

In general, there are many different polymorphisms in genes invoved in folate transport or metabolism that have been identified and studied for association with NTDs risk, but data are not yet convincing for most of them. The only one well characterized genetic risk factor is the *MTHFR* C677T. Many studies are biased by small sample size and insufficient statistical power. It is clear that other genes not related to folate pathway may play an important role in the pathogenesis of NTDs.

Studies in animal models were instrumental in dissecting the molecular pathways underlying the two major cellular events during normal neurulation: CE and apical constriction. The identification of defective genes in mouse mutants provides an entry point for the investigation of homologs and/or orthologs involved in NTDs in human beings (Zohn and Sarkar, 2008). To date, about 250 naturally occurring or experimentally induced mouse mutants with NTDs have been described, providing researchers with important tools for gene identification studies in NTDs (Harris and Juriloff, 2010) A long list of candidate genes derived from animal studies are involved in a wide range of functions at all steps of neural tube formation and patterning. The Planar Cell Polarity (PCP) pathway is an highly conserved, non-canonical Wnt-Frizzled-Dishevelled signaling cascade, that plays a key role in establishing and maintaining a coordinated polarized orientation of the cells necessary for numerous developmental procedures, including directional movements during vertebrate gastrulation and neurulation (Strutt 2003; Klein and Mlodzik 2005). PCP depends on the asymmetrical proximal-distal distribution of cell-membrane associated proteins. Genetic studies of mutants affecting complex structures in *Drosophila* have identified the genes for these asymmetrically located proteins referred to as "core PCP" components, that include transmembrane proteins such as Frizzled (Fz), Strabismus/Van Gogh (Stbm/Vang) and Flamingo (Fmi), as well as cytoplasmic proteins, including Dishevelled (Dsh/Dvl), Prickle (Pk), and Diego (Dgo) (Jenny and Mlodzik, 2006; Jones and Chen, 2007; Simons and Mlodzik, 2008). The signaling from core proteins works through a set of tissue-specific effectors to alter the cytoskeleton and gene transcription to produce morphological changes of the cells. The direction of polarization is oriented via a third PCP component, the Fat/Dachsous (Ft/Ds) system that responds to various morphogens (Goodrich and Strutt, 2011; Gray et al. 2011). PCP proteins are conserved from flies to vertebrates, and for many of the PCP genes, humans and mice have multiple homologs of the single *Drosophila* gene. A role of the PCP pathway in vertebrate cell movements in the complex morphogenetic process of CE during gastrulation and neurulation was first shown in *Xenopus* and Zebrafish. CE reflects a polarized cellular movement of mesoderm and neural plate cells from the sides of the embryo toward the midline, intercalating between the previous midline cells resulting, finally, in narrowing and lengthening of the embryonic axis during gastrulation and of the neural plate during neurulation to form a closed structure like the neural tube (Wallingford and Harland, 2002; Wang et al., 2007; Ybot-Gonzales et al., 2007). Mutations in PCP genes lead to a wide range of developmental defects, mainly a shortened body axis, a widened neural plate, and consecutively NTDs (Simons and Mlodzik, 2008).

An initial link between PCP and NTDs in mammals emerged from the discovery that the embryos of "*Loop-tail*" (*Lp*) mutant mice, developing craniorachischisis in homozygous status, showed a mutation in a "core" PCP gene, homolog to the fly Stbm/Vang, called *Vangl2* (Kibar et al., 2001, Murdoch et al., 2001). *Vangl2* presents another homolog in vertebrates, *Vangl1*, that shares 68% identity, overall similar structure and genetically interacts with *Vangl2*. (Torban et al., 2008). The human homolog *VANGL1* gene has been sequenced in a cohort of 810 NTD patients and eight missense mutations were identified both in familial and sporadic cases (Kibar et al., 2007b; Kibar et al., 2009). Sequencing of *VANGL2* in the same cohort of patients led to the identification of six novel heterozygous missense mutations in seven patients, that could be pathogenic based on genetic and initial validation data. *VANGL1* and *VANGL2* genes account only for a mutational rate of 1% on the total of NTDs patients examined. Probably they play just a partial role in the pathogenesis of NTDs, and alterations in more genes are responsible for this congenital malformation. These findings encouraged the researchers to extend the investigation of the genetic basis of NTDs to other PCP genes that collaborate with *VANGLs* genes in CE during development, to further pin down their mechanism of action.

The phenotype of PCP mice mutants offers a context for interpretation of results on PCP gene variants in human NTDs cases. Mutant mice carrying mutations in PCP genes that fail to complete neural tube closure have been described; they include: mutants of core PCP genes (*Vangl*, *Celsr*, *Dvl, Fzd, Invs*, *Prickle*), mutants of tissue specific PCP effectors (*Fuz* and *Intu*), mutants of PCP related genes (*Ptk7*, Cdx1/*Cdx2*, *Grhl3, Scrib*, *Sec24b*, *Smurfs* and *Dact1*) (Curtin et al. 2003; LiJam et al. 1997; Wang et al. 2005; Etheridge et al. 2008; Yu et al. 2010; Yokoyama et al. 1993; Tao et al. 2009; Tao et al. 2012; Kibar et al. 2001; Heydeck et al. 2009; Zeng et al. 2010; Lu et al. 2004; Savory et al. 2011; Yamamoto et al. 2008; Murdoch et al. 2001; Merte et al. 2010; Narimatsu et al. 2009; Suriben et al. 2009). Some PCP homologs when mutated cause only exencephaly, but craniorachischisis is the predominant NTD phenotype. Digenic mutants have also exencephaly and SB (Juriloff and Harris, 2012).

A review of Juriloff and Harris (2012) summarized the cumulative results of studies on human PCP genes in a variety of NTDs according to the embryological classification of NTDs made by Tortori-Donati et al. (2000). Among over 500 cases with defects of primary neurulation, including open or closed types, a total of 44 rare potentially deleterious variants absent in controls were detected among eight PCP genes (*CELSR1*, *FUZ*, *FZD6*, *PRICKLE1*, *VANGL1*, *VANGL2*, *SCRIB*, and *DACT1*) in 44 cases. Among the 44 variants, 21 have been confirmed as damaging by *in vitro* and *in vivo* assays. Deleterious genetic variants of PCP genes were all in open NTDs (craniorachischisis, myelomeningocele, anencephaly). Among defects of primary neurulation, the strongest association was found with craniorachischisis with over 20% of cases presenting mutations in PCP genes. Fewer cases with defects in secondary neurulation were studied. Among over 50 cases, five rare potentially deleterious variants in three PCP genes, *CELSR1*, *VANGL1* and *VANGL2* were found in cases with tight filum terminale, fibrolipoma or lumbar myelocistocele. Some of the studies included caudal agenesis/caudal regression and diastematomyelia. These defects arise as consequence of an anomaly of gastrulation, not a defect in neurulation. Seven potentially deleterious rare variants were found across PCP genes in caudal agenesis cases, *CELSR1*, *FZD6*, *PRICKLE1*, *VANGL1*, and *FUZ* and variants were found in two genes (*PRICKLE1* and *VANGL2*) in two diastematomyelia cases. When the parents were available, the mutations was inherited from an unaffected parent, demonstrating incomplete penetrance of the mutations. Very recently a role for *DVL2*, a pivotal multifunctional protein that is involved in both the canonical Wnt signaling and the noncanonical PCP pathway, has been demonstrated in the pathogenesis of both open and closed forms of NTDs, as reported for other PCP genes, further supporting a common mechanism involving defective PCP genes in the onset of both NTD forms (De Marco et al. 2012).

In the future researchers should explore other types of PCP gene variants such as regulatory mutations and low frequency (1 to 5%) deleterious polymorphisms. Moreover, potential genetic interactions of mutations in more than one PCP gene represent a further step forward to our understanding of the intricate genetic puzzle underlying these complex malformations.

References

Bassuk AG, Craig D, Jalali A, Mukhopadhyay A, Kim F, Charrow J, Gulbu U, Epstein LG, Bowman R, McLone D, Yagi H, Matsuoka R, Stephan DA, Kessler JA. (2005). The genetics of tethered cord syndrome. *Am J Med Genet* A 132(4):450-453.

Benjamin J, Levine J, Fux M, Aviv A, Levy D, Belmaker RH. (1995). Double-blind, placebo-controlled, crossover trial of inositol treatment for panic disorder. *Am J Psychiatry* 152: 1084–1086.

Berry RJ, Li Z, Erickson JD, Li S, Moore CA, Wang H, Mulinare J, Zhao P, Wong LY, Gindler J, Hong SX, Correa A. (1999). Prevention of neural-tube defects with folic acid in China. China-U.S. Collaborative Project for Neural Tube Defect Prevention. *N Engl J Med.* 341(20):1485-1490.

Blanco Munoz J, Lacasana M, Borja Aburto VH, Torres Sanchez LE, Garcia Garcia AM, Lopez Carrillo L. (2005). Socioeconomic factors and the risk of anencephaly in a Mexican population: A case-control study. Public Health Rep 120: 39–45.

Botto LD, Yang Q. (2000). 5,10-Methylenetetrahydrofolate reductase gene variants and congenital anomalies: a HuGE review. *Am J Epidemiol* 151: 862-877.

Bower C, Raymond M, Lumley J, Bury G. (1993). Trends in neural tube defects 1980-1989. *Med J Aust* 158(3):152-154.

Bowman RM, McLone DG, Grant JA, Tomita T, Ito JA. (2001). Spina bifida outcome: a 25-year prospective. *Pediatr Neurosurg* 34: 114-120.

Brender JD, Suarez L. (1990). Paternal occupation and anencephaly. *Am J Epidemiol* 131: 517–521.

Brody LC., Conley M, Cox C, Kirke PN, McKeever MP, Mills JL, Molloy AM, O'Leary VB, Parle-McDermott A, Scott JM, Swanson DA. (2002). A polymorphism, R653Q, in the trifunctional enzyme methylene tetrahydrofolate dehydrogenase/methenyltetrahydrofolatecyclohydrolase /formyltetrahydrofolate synthetase is a maternal genetic risk factor for neural tube defects: report of the Birth Defects Research Group. *Am J Hum Genet* 71: 1207-1215.

Campbell LR, Dayton DH, Sohal GS. (1986). Neural tube defects: a review of human and animal studies on the etiology of neural tube defects. *Teratology* 34(2):171-187. Review.

Canfield MA, Annegers JF, Brender JD, Cooper SP, Greenberg F. (1996). Hispanic origin and neural tube defects in Houston/Harris County, Texas. I. Descriptive epidemiology. *Am J Epidemiol* 143(1):1-11.

Canick JA, Saller DN Jr. (1993). Maternal serum screening for aneuploidy and open fetal defects. *Obstet Gynecol Clin North Am* 20(3):443-454.

Carter CO, Evans K. (1973). Spina bifida and anencephalus in grater London 1. *J Med Genet* 10(3): 209-234.

Carter CO, Evans KA, Till K. (1976). Spinal dysraphisms: genetic relation to neural tube malformations. *J Med Genet* 13: 343-350.

Catala M, Teillet MA, De Robertis EM, Le Douarin ML. (1996). A spinal cord fate map in the avian embryo: while regressing, Hensen's node lays down the notochord and floor plate thus joining the spinal cord lateral walls. *Development* 122(9):2599-2610.

Cavalli P, Tedoldi S, Riboli B. (2008). Inositol supplementation in pregnancies at risk of apparently folate-resistant NTDs. *Birth Defects Res A Clin Mol Teratol* 82: 540–542.

Cavalli P, Tonni G, Grosso E, Poggiani C. (2011). Effects of inositol supplementation in a cohort of mothers at risk of producing an NTD pregnancy. *Birth Defects Res A Clin Mol Teratol* 91(11):962-965.

Centers for Disease Control (CDC). (1992). Recommendations for the use of folic acid to reduce the number of cases of spina bifida and other neural tube defects. *MMWR* 41(No. RR-14):1-7.

Centers for Disease Control and Prevention (CDC). (2002). Folate status in women of childbearing age, by race/ethnicity--United States, 1999-2000. *MMWR Morb Mortal Wkly Rep* 51(36):808-810.

Christensen B, Arbour L, Tran P, Leclerc D, Sabbaghian N, Platt R, Gilfix BM, Rosenblatt DS, Gravel RA, Forbes P, Rozen R. (1999). Genetic polymorphisms in methylenetetrahydrofolate reductase and methionine synthase, folate levels in red blood cells, and risk of neural tube defects. *Am J Med Genet* 84: 151-157.

Colodny L, Hoffman RL. (1998). Inositol–clinical applications for exogenous use. *Altern Med Rev* 3: 432–447

Committee on Developmental Toxicology, National Research Council. (2000). Scientific Frontiers in Developmental Toxicology and Risk Assessment. National Academic Press, Washington, DC.

Copp AJ, Brook FA. (1989). Does lumbosacral spina bifida arise by failure of neural folding or by defective canalisation? *J Med Genet* 26: 160–166.

Copp AJ, Greene ND, Murdoch JN. (2003). The genetic basis of mammalian neurulation. *Nat Rev Genet* 4: 784–793.

Correa, A., Stolley, A., and Liu, Y. (2000). Prenatal tea consumption and risks of anencephaly and spina bifida. *Ann Epidemiol* 10, 476–477.

Curtin JA,Quint E,Tsipouri V, *et al.* (2003). Mutation of Celsr1 disrupts planar polarity of inner ear hair cells and causes severe neural tube defects in the mouse. *Curr Biol* 13: 1129–1133.

Czeizel AE, Dudas I. (1992). Prevention of the first occurence of neural-tube defects by periconceptional vitamin supllementation. *N Engl J Med* 327, 1832-1835.

De Marco P, Calevo MG, Moroni A, Arata L, Merello E, Finnell RH, Zhu H, Andreussi L, Cama A, Capra V. (2002). Study of MTHFR and MS polymorphisms as risk factors for NTD in the Italian population. *J Hum Genet* 47:319-324.

De Marco P, Merello E, Calevo MG, Mascelli S, Raso A, Cama A, Capra V. (2006). Evaluation of methylenetetrhydrofolate-dehydrogenase 1958G>A polymorphism for neural tube defect risk. *J Hum Genet* 51:98-103.

De Marco P, Merello E, Cama A, Kibar Z, Capra V. (2011). Human neural tube defects: genetic causes and prevention. *Biofactors*.37(4):261-268.

De Marco P, Merello E, Consales A, Piatelli G, Cama A, Kibar Z, Capra V. (2012). Genetic Analysis of Disheveled 2 and Disheveled 3 in Human Neural Tube Defects. *J Mol Neurosci.* [Epub ahead of print]

Dietl J. (2005). Maternal obesity and complications during pregnancy. *J Perinat Med* 33: 100–105.

Dudin A. (1997). Neural tube defect among Palestinians: a hospital-based. *Ann Trop Paediatr* 17: 217-222.

Elmore CL, Wu X, Leclerc D, Watson ED, Bottiglieri T, Krupenko NI, Krupenko SA, Cross JC, Rozen R, Gravel RA, Matthews RG. (2007). Metabolic derangement of methionine and folate metabolism in mice deficient in methionine synthase reductase. *Mol Genet Metab* 91:85-97.

Elwood JM, Elwood JH. (1980). *Epidemiology of anencephalus and spina bifida*. New York, Toronto, Oxford University Press, 85-119.

Elwood JM, Little J, Elwood JH. (1992). Epidemiology and control of Neural Tube Defects. Oxford:Oxford University Press.

Ethen MK, Canfield MA. (2002). Impact of including elective pregnancy terminations before 20 weeks gestation on birth defect rates. *Teratology* 66 Suppl 1:S32-35

EUROCAT Working Group. (1991). Prevalence of neural tube defects in 20 regions of Europe and the impact of prenatal diagnosis 1980-1986. *J Epidemiol Commun Health* 45: 52-58.

EUROCAT Working Group. (1997). Report 7:15 years of Surveillance of Congenital Anomalies in Europe, 1980-1994. Brussels: Scientific Institute of Public Helath-Louis Pasteur.

EUROCAT Working Group. (2002): Report 8: surveillance of congenital anomalies in Europe, 1980-1999. Newtownabbey, Northern Ireland. EUROCAT Central Registry , University of Ulster.

Fleming A, Copp AJ. (2000). A genetic risk factor for mouse neural tube defects: defining the embryonic basis. *Hum Mol Genet* 9: 575–581.

Frosst P, Blom HJ, Milos R, Goyette P, Sheppard CA, Matthews RG, Boers GJH, den Heijer M, Kluijtmans LAJ, van den Heuvel LP, Rozen R. (1995). A candidate genetic risk for vascular disease: a common mutation in methylenetetrahydrofolate reductase. *Nat Genet* 10:111–113.

Gonzales-Herrera L, Garcia-Escalante G, Castillo-Zapata I, Canto-Herrera J, Caballos-Quintal J, Pinto-Escalante D, Diaz-Rubio F, Del Angel RM, Orozco-Orozco L. (2002). Frequency of the thermolabile vatiant C677T in the *MTHFR* gene and lack of association with neural tube defects in the State of Yucatan, Mexico. *Clin Genet* 62: 394-398.

Goodrich LV, Strutt D. (2011). Principles of planar polarity in animal development. *Development* 138: 1877-1892.

Gray RS, Roszko I, Solnica-Krezel L. (2011). Planar cell polarity: coordinating morphogenetic cell behaviors with embryonic polarity. *Dev Cell* 21: 120-133.

Greene ND, Copp AJ. (2009). Development of the vertebrate central nervous system: formation of the neural tube. *Prenat Diagn*. 29(4):303-11.

Harris and Juriloff. (2010). An update to the list of mouse mutants with neural tube closure defects and advances toward a complete genetic perspective of neural tube closure. *Birth Defects Res A Clin Mol Teratol* 88(8):653-669.

Hendricks KA, Simpson JS, Larsen RD. (1999). Neural tube defects along the Texas-Mexico border, 1993-1995. *Am J Epidemiol* 149 (12):1119-1127.

Heydeck W,Zeng H,Liu A. (2009). Planar cell polarity effector gene Fuzzy regulates cilia formation and Hedgehog signal transduction in mouse. *Dev Dyn* 238: 3035–3042.

Hibbard BM: (1964). The role of folic acid in pregnancy with particular reference to anaemia abruption and abortion. *J Obstet Gynaecol Br Commonw*. 71:529-542.

Howlett A,Ohlsson A. (2003). Inositol for respiratory distress syndrome in preterm infants. *Cochrane Database Syst Rev* 4: CD000366.

Jenny A, Mlodzik M. (2006). Planar cell polarity signaling: a common mechanism for cellular polarization. *Mt Sinai J Med* 73(5):738-50.

Jones C, Chen P. (2007). Planar cell polarity signaling in vertebrates. *Bioessays* 29(2):120-32.

Khoury MJ, Moore CA, James LM, Cordero JF. (1992). The interaction between dsymorphology and epidemiology: methodologic issues of lumping and splitting. *Teratology* 45: 133-138.

Kibar Z, Vogan KJ, Groulx N, Justice MJ, Underhill DA, Gros P. (2001). Ltap, a mammalian homolog of Drosophila Strabismus/Van Gogh, is altered in the mouse neural tube mutant Loop-tail. *Nat Genet* 28(3):251-255.

Kibar Z, Capra V, Gros P. (2007a). Toward understanding the genetic basis of neural tube defects. *Clin Genet* 71(4):295-310.

Kibar Z, Torban E, McDearmid JR, Reynolds A, Berghout J, Mathieu M, Kirillova I, De Marco P, Merello E, Hayes JM, Wallingford JB, Drapeau P, Capra V, Gros P. (2007b) Mutations in VANGL1 associated with neural-tube defects. *N Engl J Med* 356 (14):1432-1437.

Kibar Z, Bosoi CM, Kooistra M, Salem S, Finnell RH, De Marco P, Merello E, Bassuk AG, Capra V, Gros P. (2009). Novel mutations in VANGL1 in neural tube defects. *Hum Mutat* 30(7):E706-715.

Klein TJ, Mlodzik M. (2005). Planar cell polarization: an emerging model points in the right direction. *Annu Rev Cell Dev Biol* 21:155-176.

Koch MC, Stegmann K, Ziegler A, Schroter B, Ermert A. (1998). Evaluation of the *MTHFR* C677T allele and the *MTHFR* gene locus in a German spina bifida population. *Eur J Pediatr* 157: 487-492.

ICBDMS. International Clearinghouse for Birth Defects Monitoring Systems. (1993). Annual Report 1993 (with data for 1991). Rome: International Centre for Birth Defects.

ICBDMS. International Clearinghouse for Birth Defects Monitoring Systems. (1994). Annual Report 1994 (with data for 1992). Rome: International Centre for Birth Defects.

ICBDMS. International Clearinghouse for Birth Defects Monitoring Systems. (1995). Annual Report 1995 (with data for 1993). Rome: International Centre for Birth Defects.

ICBDMS. International Clearinghouse for Birth Defects Monitoring Systems. (1996). Annual Report 1996 (with data for 1994). Rome: International Centre for Birth Defects.

ICBDMS. International Clearinghouse for Birth Defects Monitoring Systems. (1997). Annual Report 1997 (with data for 1995). Rome: International Centre for Birth Defects.

ICBDMS. International Clearinghouse for Birth Defects Monitoring Systems. (1998). Annual Report 1998 (with data for 1996). Rome: International Centre for Birth Defects.

ICBDMS. International Clearinghouse for Birth Defects Monitoring Systems. (1999). Annual Report 1999 (with data for 1997). Rome: International Centre for Birth Defects.

ICBDMS. International Clearinghouse for Birth Defects Monitoring Systems. (2000). Annual Report 2000 (with data for 1998). Rome: International Centre for Birth Defects.

ICBDMS. International Clearinghouse for Birth Defects Monitoring Systems. (2001). Annual Report 2001 (with data for 1999). Rome: International Centre for Birth Defects.

ICBDMS. International Clearinghouse for Birth Defects Monitoring Systems. (2002). Annual Report 2002 (with Data for 2000). Rome: International Centre for Birth Defects.

Laurence KM, James N, Miller MH, Tennant GB, Campbell H. (1981). Double-blind randomised controlled trial of folate treatment before conception to prevent recurrence of neural-tube defects. *Br Med J* (Clin Res Ed). 282(6275):1509-1511.

LecK I. (1996). Changes in the incidence of neural tube defects. *Lancet* 2: 791-793.

Levine J, Aviram A, Holan A, Ring A, Barak Y, Belmaker RH. (1997). Inositol treatment of autism. *J Neural Transm* 104: 307–310.

Little J, Elwood H. (1992). Socio-economic status and occupation. *In*: Elwood JM LJ, Elwood H, *editors*. Epidemiology and Control of Neural Tube Defects. Oxford: Oxford University Press: 456-520.

Loeken, M. R. (2005). Current perspectives on the causes of Neural Tube Defects resulting from diabetic pregnancy. *Am. J. Med. Genet. Part C* 135C, 77–87.

Marini NJ, Hoffmann TJ, Lammer EJ, Hardin J, Lazaruk K, Stein JB, Gilbert DA, Wright C, Lipzen A, Pennacchio LA, Carmichael SL, Witte JS, Shaw GM, Rine J. (2011). A genetic signature of spina bifida risk from pathway-informed comprehensive gene-variant analysis. *PLoS One* 6(11):e28408.

Mastroiacovo P, Leoncini E. (2011). More folic acid, the five questions: why, who, whe, how much, and how. *Biofactors* 37(4):272-279.

Matalon S., Schechtman S., Goldzweig G., Ornoy A. (2002). The teratogenic effect of carbamazepine: A meta-analysis of 1255 exposures. *Reprod. Toxicol.* 5, 33–39.

Matte TD, Mulinare J, Erickson JD. (1993). Case-control study of congenital defects and parental employment in health care. *Am J Ind Med* 24: 11–23.

McDonnell RJ, Johnson Z, Delaney V, Dack P. (1999). East Ireland 1980-1994: epidemiology of neural tube defects. *J Epidemiol Community Health* 53(12):782-788.

Merte J,Jensen D,Wright K, *et al.* (2010). Sec24b selectively sorts Vangl2 to regulate planar cell polarity during neural tube closure. *Nat Cell Biol* 12:41–46; *sup pp* 1–8.

Moretti, M. E., Bar-Oz, B., Fried, S., and Koren, G. (2005). Maternal hyperthermia and the risk for neural tube defects in offspring: systematic review and meta-analysis. *Epidemiology* 16, 216–219.

Morrison K, Papapetrou C, Hol FA, Mariman EC, Lynch SA, Burn J, Edwards YH. (1998). Susceptibility to spina bifida; an association study of five candidate genes. *Ann Hum Genet* 62:379-396.

MRC Vitamin Study Research Group. *(1991)* Prevention of neural tube defects: results of the Medical Research Council Vitamin Study. *Lancet* 338:131-137.

Murdoch JN, Doudney K, Paternotte C, Copp AJ, Stanier P. (2001). Severe neural tube defects in the loop-tail mouse result from mutation of Lpp1, a novel gene involved in floor plate specification. *Hum Mol Genet* 10(22):2593-601.

Myrianthopoulos NC and Melnick M. (1987). Studies in neural tube defects. I. Epidemiological and etiologic aspects. *Am J Med Genet* 26, 783–796.

Nakano KK. (1993). Anencephaly: A review. *Developmental Medicine and Child Neurology* 15(3): 383-400.

Nakatsu T, Uwabe C, Shiota K. (2000). Neural tube closure in humans initiates at multiple sites: evidence from human embryos and implications for the pathogenesis of neural tube defects. *Anat Embryol* 201: 455–466.

Narimatsu M ,Bose R, Pye M, Zhang L, Miller B, Ching P, Sakuma R, Luga V, Roncari L, Attisano L, Wrana JL. *(*2009). Regulation of planar cell polarity by Smurf ubiquitin ligases. *Cell* 137: 295–307.

National Health & Medical Research Council. (1994). Revised statement on the relationship between dietary folic acid and neural tube defects such as spina bifida. *Journal of Paediatrics and Child Health* 30:476–477.

Nestler JE, Jakubowicz DJ, Reamer P, Gunn RD, Allan G. *(*1999). Ovulatory and metabolic effects of D-chiro-inositol in the polycystic ovary syndrome. *N Engl J Med* 340: 1314–1320.

O'Leary VB, Mills JL, Parle-McDermott A, Pangilinan F, Molly AM, Cox C, Weiler A, Conley M, Kirke PN, Scott JM, Brody LC. (2005). Screening for new *MTHFR* polymorphisms and NTD risk. *Am J Med Genet* 38:99-106.

O'Reilly GC, Shields LE. (2000) Karyotyping for isolated neural tube defects. A report of two cases.*J Reprod Med* 45(11):950-952.

O'Rahilly R, Muller F. (2002). The two sites of fusion of the neural folds and the two neuropores in the human embryo. *Teratology* 65:162–170.

Owen TJ, Halliday JL, Stone CA. (2000). Neural tube defects in Victoria, Australia: potential contributing factors and public health implications. *Aust N Z J Public Health* 24(6):584-589.

Ou CY, Stevenson RE, Brown VK, Schwart CE, Allen WP, Khoury MJ, Rozen R, Oakley GP, Adams MJ. (1996). Methylenetetrahydrofolate reductase genetic polymorphism as a risk factor for neural tube defects. *Am J Med Genet* 63: 610-614.

Palatnik A, Frolov K, Fux M, Benjamin J. (2001) Double-blind, controlled, crossover trial of inositol versus fluvoxamine for the treatment of panic disorder. *J Clin Psychopharmacol* 21: 335–339.

Partington MD, McLone DG. (1995). Hereditary factors in the etiology of neural tube defects. Results of a survey. *Pediatr Neurosurg* 23, 311-316.

Pietrzik K, Lamers Y, Bramswing S and Prinz-Langerohl R. (2007). Calculation of red blood cell folate steady conditions and elimination kinetics after daily supplementation with various folate forms and doses in women of childbearing age. *Am J Clin Nutr* 86: 1414-1419.

Purves D, Lichtman JW. 1985. Principles of neural development. Sunderland, MA: Sinauer

Rampersaud E, Bassuk AG, Enterline DS, George TM, Siegel DG, Melvin EC, Aben J, Allen J, Aylsworth A, Brei T, Bodurtha J, Buran C, Floyd LE, Hammock P, Iskandar B, Ito J, Kessler JA, Lasarsky N, Mack P, Mackey J, McLone D, Meeropol E, Mehltretter L, Mitchell LE, Oakes WJ, Nye JS, Powell C, Sawin K, Stevenson R, Walker M, West SG, Worley G, Gilbert JR, Speer MC. (2005). Whole genomewide linkage screen for neural tube defects reveals regions of interest on chromosomes 7 and 10. *J Med Genet* 42 (12):940-946.

Rasmussen SA, Olney RS, Holmes LB, Lin AE, Keppler_Noreuil KM, Moore CA. (2003). Guidelines for case classification for the National Birth Defects Prevention Study, Birth *Defects Res PartA* 67: 193-201.

Ray JG, Vermeulen MJ, Meier C, Wyatt PR. (2004).Risk of congenital anomalies detected during antenatal serum screening in women with pregestational diabetes. *QJM* 97: 651–653.

Revilla JIG, Hernandez FN, Martin MTC, Salvador MT, Romero JG. (2003). C677T and A1298C *MTHFR* polymorphisms in the etiology of neural tube defects in Spanish population. *Med Clin* 120: 441-445.

Risch N. (1990). Linkage strategies for genetically complex traits. II. The power of affected relative pairs. *Am J Hum* Genet 46, 229-241.

Rossi A, Biancheri R, Cama A, Piatelli G, Ravegnani M, Tortori-Donati P. (2004). Imaging in spine and spinal cord malformations. *Eur J Radiol* 50(2):177-200.

Sadler TW. (1998). Mechanisms of neural tube closure and defects. *Ment Retard Dev Disabil Res* 4: 247-253.

Savory JG, Mansfield M, Rijli FM, Lohnes D. (2011). Cdx mediates neural tube closure through transcriptional regulation of the planar cell polarity gene Ptk7. *Development* 138: 1361–1370.

Schmidt RJ, Romitti PA., Burns TL., Browne ML, Drushel CM, Oln RS, Birth Defects Prevention Study. (2009). Maternal caffeine consumption and risk of neural tube defects. *Birth Defects Res. Part A: Clin Mol Teratol* 85:879–889.

Schoenwolf GC. (1984). Histological and ultrastructural studies of secondary neurulation of mouse embryos. *Am J Anat* 169: 361–374.

Sever LE. (1995). Looking for causes of neural tube defects: Where does the environment fit in? *Environ Health Perspect* 103 (Suppl. 6): 165–171.

Shaw GM, Jensvold NG, Wasserman CR, Lammer EJ. (1994) Epidemiologic characteristics of phenotypically distinct neural tube defects among 0.7 million California births, 1983-1987. *Teratology*.49(2):143-149.

Shaw GM, Velie EM., Schaffer D. (1996). Risk of neural tube defect-affected pregnancies among obese women. *JAMA* 275, 1093–1096.

Shaw GM, Nelson V, Olshan AF. (2002). Paternal occupational group and risk of offspring with neural tube defects. *Paediatr Perinat Epidemiol* 16: 328–333.

Shaw GM, Ouach T, Nelson V, Carmichael SL, Schaffer DM, Selvin S, Yang W. (2003). Neural tube defects associated with maternal periconceptional dietary intake of simple sugars and glycemic index. *Am. J. Clin. Nutr*. 78, 972–978.

Shenefelt RE. (1972). Morphogenesis of malformations in hamsters caused by retinoic acid: Relation to dose and stage at treatment. *Teratology* 5: 103–118.

Shepard TH, Brent RL, Friedman JM, Jones KL, Miller RK, Moore CA, Polifka JE. (2002). Update on new developments in the study of human teratogens. *Teratology* 65: 153–161.

Simons M, Mlodzik M. (2008). Planar cell polarity signaling: from fly development to human disease. *Annu Rev Genet* 42:517-540.

Smith JL, Schoenwolf GC. (1997). Neurulation: coming to closure. *Trends Neurosci*. 20(11):510-517.

Smithells RW, Sheppard S, Schorah CJ. (1976). Vitamin dificiencies and neural tube defects. *Arch Dis Child.* 51(12):944-950.

Smithells RW, Sheppard S. (1980a) Possible prevention of neural-tube defects by periconceptional vitamin supplementation. *Lancet* 1(8169):647.

Smithells RW, Sheppard S, Schorah CJ, Seller MJ, Nevin NC, Harris R, Read AP, Fielding DW. (1980b) Possible prevention of neural-tube defects by periconceptional vitamin supplementation. *Lancet* 1(8164):339-340.

Smithells RW, Nevin NC, Seller MJ, Sheppard S, Harris R, Read AP, Fielding DW, Walker S, Schorah CJ, Wild J. *(1983). Further experience of vitamin supplementation for prevention of neural tube defect recurrences. Lancet* 1:1027–1031.

Stamm DS, Rampersaud E, Slifer SH, Mehltretter L, Siegel DG, Xie J, Hu-Lince D, Craig DW, Stephan DA, George TM, Gilbert JR, Speer MC; NTD Collaborative Group. (2006). High-density single nucleotide polymorphism screen in a large multiplex neural tube defect family refines linkage to loci at 7p21.1-pter and 2q33.1-q35. *Birth Defects Res A Clin Mol Teratol* 76(6):499-505.

Stegmann A. Ziegler ET, Ngo N, Kohlschmidt B, Schröter, A. Ermert, M.C. Koch. (1999). Linkage disequilibrium of MTHFR genotypes 677C/T-1298A/C in the German population and association studies in probands with neural tube defects (NTD) *Am J Med Genet.* 87:23–29

Suriben R, Kivimäe S, Fisher DA, Moon RT, Cheyette BN. (2009). Posterior malformations in Dact1 mutant mice arise through misregulated Vangl2 at the primitive streak. *Nat Genet* 41: 977–985.

Swanson DA, Liu ML, Baker PJ, Garrett L, Stitzel M, Wu J, Harris M, Banerjee R, Shane B, Brody LC. (2001). Targeted disruption of the methionine synthase gene in mice. *Mol Cell Biol* 21:1058-1065.

Strutt D. (2003). Frizzled signalling and cell polarisation in Drosophila and vertebrates. *Development* 130(19):4501-13.

Tao H, Suzuki M, Kiyonari H, Abe T, Sasaoka T, Ueno N. (2009). Mouse prickle1, the homolog of a PCP gene, is essential for epiblast apical-basal polarity. *Proc Natl Acad Sci USA* 106: 14426–14431.

Tao H, Inoue KI, Kiyonari H, Bassuk AG, Axelrod JD, Sasaki H, Aizawa S, Ueno N. *(*2012). Nuclear localization of Prickle2 is required to establish cell polarity during early mouse embryogenesis. *Dev Biol* 15: 138–148.

Torban E, Patenaude AM, Leclerc S, Rakowiecki S, Gauthier S, Andelfinger G, Epstein DJ, Gros P. (2008). Genetic interaction between members of the Vangl family causes neural tube defects in mice. *Proc Natl Acad Sci USA.* 105(9):3449-3454.

Tortori-Donati P, Rossi A,Cama A. *(*2000). Spinal dysraphism: a review of neuroradiological features with embryological correlations and proposal for a new classification. *Neuroradiology* 42: 471–491.

Van Allen MI, Kalousek DK, Chernoff GF, Juriloff D, Harris M, McGillivray BC, Yong SL, Langlois S, MacLeod PM, Chitayat D. (1993). Evidence for multi-site closure of the neural tube in humans. *Am J Med Genet* 47: 723–743.

Van der Put NM, Steegers-Theunissen RP, Frosst P, Trijbels FJ, Eskes TK, van den Heuvel LP, Mariman EC, den Heyer M, Rozen R, Blom HJ. (1995). Mutated methylenetetrahydrofolate reductase as risk factor for spina bifida. Lancet 346, 1070-1071.

van der Put NM, Gabreels F, Stevens EM, Smeitink JA, Trijbels FJ, Eskes TK, van den Heuvel LP, Blom HJ. (1998). A second common mutation in the methylenetetrahydrofolate reductase gene: an additional risk factor for neural-tube defects? *Am J Hum Genet* 62, 1044-1051.

Vintzileos AM, Campbell WA, Nochimson DJ, Weinbaum PJ. (1987). Antenatal evaluation and management of ultrasonically detected fetal anomalies. *Obstet Gynecol* 69(4):640-660.

Wallingford JB, Harland RM. (2002). Neural tube closure requires Dishevelled-dependent convergent extension of the midline. *Development.* Dec;129(24):5815-5825.

Wang J, Mark S, Zhang X, Qian D, Yoo SJ, Radde-Gallwitz K, Zhang Y, Lin X, Collazo A, Wynshaw-Boris A, Chen P. (2005). Regulation of polarized extension and planar cell polarity in the cochlea by the vertebrate PCP pathway. *Nat Genet* 37: 980–985.

Wang Y, Nathans J. (2007). Tissue/planar cell polarity in vertebrates: new insights and new questions. *Development* 134, 647–658.

Wasserman CR, Shaw GM, Selvin S, Gould JB, Syme SL. (1998). Socioeconomic status, neighborhood social conditions, and neural tube defects. *Am J Public Health* 88: 1674–1680.

Weisberg I, Tran P, Christensen B, Sibani S, Rozen R. (1998). A second genetic polymorphism in methylenetetrahydrofolate reductase (*MTHFR*) associated with decreased enzyme activity. *Mol Genet Metab* 64: 169-172.

Whiteman D, Murphy M, Hey K, O'Donnell M, Goldacre M. (2000). Reproductive factors, subfertility, and risk of neural tube defects: a case-control study based on the Oxford Record Linkage Study Register. *Am J Epidemiol* 152(9):823-828.

Wilson JG. (1973). Environment and Birth Defects. Academic Press, New York.

Wilson A, Platt R, Wu Q, Leclerc D, Christensen B, Yang H, Gravel RA, Rozen R. (1999). A common variant in methionine synthase reductase combined with low cobalamin (vitamin B12) increases risk for spina bifida. *Mol Genet Metab* 67, 317–323.

Yamamoto S,Nishimura O,Misaki K, Nishita M, Minami Y, Yonemura S, Tarui H, Sasaki H. (2008). Cthrc1 selectively activates the planar cell polarity pathway of Wnt signaling by stabilizing the Wnt-receptor complex. *Dev Cell* 15: 23–36.

Yen IH, Khoury MJ, Erickson JD, James LM, Waters GD, Berry RJ. (1992). The changing epidemiology of neural tube defects. United States, 1968-1989. *Am J Dis Child* 146(7):857-861.

Ybot-Gonzalez P, Savery D, Gerrelli D, Signore M, Mitchell CE, Faux CH, Greene ND, Copp AJ. (2007). Convergent extension, planar-cell-polarity signalling and initiation of mouse neural tube closure. *Development* 134(4):789-799.

Yokoyama T, Copeland NG, Jenkins NA, Montgomery CA, Elder FF, Overbeek PA. (1993). Reversal of left-right asymmetry: a situs inversus mutation. *Science* 260: 679–682.

Yu H, Smallwood PM, Wang Y, Vidaltamayo R, Reed R, Nathans J. (2010). Frizzled 1 and frizzled 2 genes function in palate, ventricular septum and neural tube closure: general implications for tissue fusion processes. *Development* 137: 3707–3717.

Zeng H, Hoover AN, Liu A. (2010). PCP effector gene Inturned is an important regulator of cilia formation and embryonic development in mammals. *Dev Biol* 339: 418–428

Zohn IE, Sarkar AA. (2008). Modeling neural tube defects in the mouse. *Curr Top Dev Biol* 84:1-35.

In: Neural Tube Defects
Editor: Alberich Klein

ISBN: 978-1-62417-892-4

Chapter II

Primary Prevention of Neural Tube Defects: Not Only Preconception Folic Acid Supplementation

Carmen De Luca, Ilenia Mappa, Angelo Licameli, Giovanni De Luca and Marco De Santis
Department of Obstetrics and Gynecology, Policlinico "A.Gemelli", Università Cattolica del Sacro Cuore, Rome, Italy

Abstract

Neural tube defects (NTDs) are birth anomalies resulting from the embryonic neural tube's failure to fuse near the 28th day after conception. Several studies have demonstrated that periconception Folic Acid (FA) supplementation decreases the risk of NTDs by about 70 %. As a result, is recommended to consume 0,4 mg folic acid per day during the periconception period to reduce the risk of first occurrence NTDs. But several maternal conditions, increasing the risk of congenital NTDs, need a preconception evaluation and an appropriate intervention. It's well known that certain medications (such as antiepileptics and other anti-folate drugs) can to increase the risk of NTDs, as well as alterations in maternal folate and homocysteine metabolism. An augmented risk has been also documented in maternal diabetes and obesity, probably due to a sustained state of hyperglycemia and/or hyperinsulinemia. In addition, some studies have suggested that periconception maternal smoking can increase the risk of NTDs in offspring. Women suffering from malabsorption syndrome also represent a high-risk population, such as patient affected by NTDs or with previous pregnancies with NTDs. Some studies have demonstrated that maternal occupational exposure to organic compounds during early pregnancy can also represent an important risk factor. In conclusion, primary prevention of NTDs not only includes preconception FA supplementation but requires an adequate multidisciplinary risk assessment.

Introduction

Neural tube defects (NTDs) are birth anomalies resulting from the embryonic neural tube's failure to fuse near the 28th day after conception. NTDs include anencephaly, spina bifida and encephalocele. It has been estimated that each year there are approximately 300,000 new cases of neural tube defects [1]. In 1998, the United States and Canada implemented mandatory Folic Acid (FA) fortification to reduce NTDs. Noting the reduction of NTD rates in the United States (U.S.) and Canada following mandatory FA fortification, several other countries have successfully instituted or are currently considering FA fortification [2-3]. In the United States the incidence is of 1.4–1.6/1,000 live births [4] and the number of pregnancy affected by NTDs declined after fortification of enriched cereal grain products with folic acid from 4000 (1995-1996) to 3,000 pregnancies (1999-2000) [5].

European incidence is about 4500 cases per year [6]. Various genetic and environmental factors are related to NTD. Several studies have demonstrated that periconception FA supplementation decreases the risk of NTDs by about 70 % [7-8]. In 1992, the U.S. Public Health Service recommended that "all women of childbearing age in the United States who are capable of becoming pregnant should consume 0.4 mg of folic acid per day for the purpose of reducing their risk of having a pregnancy affected with spina bifida or other NTDs" [9]. But this dose cannot be sufficient for some categories of women who have an augmented risk of NDTs and might benefit for higher doses before conception and during the first trimester of pregnancy. The American Congress of Obstetricians and Gynecologists (ACOG) suggested that folic acid supplementation of 0,4 mg/day is recommended for low-risk women, and folic acid supplementation of 4mg/day is recommended for women at high risk for NTD [10].

Previous Pregnancy with NDT

In 95% of cases, NTDs occur spontaneously in women with no family history but in 5% of NTDs occur in women with a positive family history. If there is one previous affected pregnancy, the recurrence risk increases to 10-fold while this risk doubles for two previous affected pregnancies and quadruples for three previous affected pregnancies [11]. FA supplementation can reduce the risk of recurrence by at least 85% [12].

According to The Centers for Disease Control and Prevention (CDC), women of childbearing age who have had a previous NTD-affected pregnancy should take 0.4 mg of FA daily while not pregnant and they should increase the amount of FA consume to 4.0 mg daily when they are planning a pregnancy, from 1 month before conception through the first 3 months of pregnancy [13].

The reduction in risk among these women who took FA before becoming pregnant was estimated to be 87% in randomized trials 85–100% in observational studies [12]. A dose-response association has been reported, with a lower risk associated with a daily dose of greater than 0.8 mg/day [14].

The effectiveness of NTD recurrence prevention programs depends on the likelihood that woman will take FA. In countries lacking a well assessed NTD recurrence prevention program, as Texas, FA supplement use among women with a prior NTD-affected pregnancy was reported to be about one-third of supplement use [15-17].

In South Carolina has been reported the highest rate of supplement. FA supplementation is promoted throughout individual counseling and direct providing or paying for folic acid supplements to individuals [17].

Hyperomocisteinemia and Polymorphisms Associated with NTDs

FA, also known as vitamin B_9, vitamin B_c or folacin, is a vitamin that is essential for DNA synthesis and methylation [18]. NTDs have been associated with several polymorphisms of enzymes involved in these critical metabolic pathways. Humans cannot produce folate, but they can introduce polyglutamate folate by diet or synthetic monoglutamate form. Alimentary sources of folate include green leafy vegetables, fruits, liver, legumes and wheat bread. Synthetic folic acid can absorb directly, while polyglutamete folate is changed into monoglutamate by folypoli-γ- glutamate carbocypeptidase II and then is converted in dyhidrofolate and tetrahydrofolate (THF). Before transfer in portal circulation, THF is transformed in 5,10-methylene-THF and 5-methyl-THF [19]. The polymorphism of the gene coding for the folypoli-γ- glutamate carbocypeptidase II results in a reduced activity of this enzyme and in a consequent reduction of folate plasma level. Moreover, the presence of autoantibodies against the folate receptors can inhibit folate uptake [20]. Polymorphism of genes involved in the cellular folate transportation, such as the reduced folate carrier gene (RFC1), has also been associated with NTDs [21]. Methylenetetrahydrofolate (MTHF) dehydrogenase is an enzyme involved in purine synthesis, providing 10-formyltetraydrofolate and in presence of polymorphism of this gene DNA damages can occur [22]. A polymorphic 19 bp deletion allele within intron-1 of dihydrofolate reductase (DHFR) has also been investigated has possible cause of NTDs [23].

Homocisteine is a toxic metabolite that requires vitamin B12 as cofactor and 5-MTHF as methyl donor to be converted in methionine. Since hyperhomicysteinemia is one of the most well established risk factor for NTDs, this fact can explain the role folate deficiency in the genesis of NTDs. Methylenetetrahydrofolate reductase (MTHFR) catalyzes the conversion of 5,10-methylenetetrahydrofolate to 5-methyltetrahydrofolate. The most important cause of hyperhomocysteinemia and/or a lack of methionine is the polymorphism of the MTHFR gene. The MTHFR C667T polymorphism results in a thermolabile form of the enzyme, with a 40% reduction in its activity of heterozygote and with a 70% reduction activity of the homozygote forms [19]. The population frequency of the C677T allele showed geographical and ethnic variations: it is higher in Italy and among Hispanics and was low among African-American and in some areas of sub-Saharan Africa. In North America, the frequency of homozygotes increased from western Canada (5.8%) to south eastern U.S. (10,7%) and peaked in Mexico (32.3%) [24]. In Europe the frequency of C677T homozygosity is about 10-12% while the frequency of the heterozygote form is 40% [25].

Maternal C677T homozygosity appears to be a risk factor (Odds ratio = 2.0; 95%, Confidence interval (CI) = 1.5- 2.8). The OR for being a mother of an affected child is 1.2 (95% CI= 0.96- 1.5) for the heterozygous C677T genotypes. The homozygote form in infants is associated with a moderately increased risk for spina bifida (OR = 1.8; 95% CI=1.4- 2.2) [24]. Another common polymorphism of the MTHFR gene, A1298C, has also been described as a risk mutation, alone or combined with the C677T allele also could be associated with an increased risk for spina bifida [24-26]

Maternal Intake of Micronutrients

It well established that several micronutrients function as enzyme cofactors or methyl-donors in folate metabolism and lack of these cofactors, increasing maternal hyperomocistysteinemia, has been associated with NTDs. Moreover, oxidative stress can increase the risk of NTS and micronutrients with antioxidant functions can reduce this risk. As a consequence, maternal nutritional status plays an important role in prevention of NTDs. Vitamin A is a fat-soluble vitamin important for visual function , cell differentiation, immune system, haematopoiesis, regulation of mucosal surfaces, intestinal iron uptake and reproduction. Vitamin A is considered teratogenic in high concentrations. A recent study demonstrated that normal-weighted women with highest intake of vitamin A have a lower risk of anencephaly (OR=0.57, 95% CI= 0.10-0.86) [27]. Vitamin B6 is a co-enzymes important in protein metabolism and in development of the central nervous system. A diet rich of vitamin B6 seems to decrease the risk of spina bifida among Hispanic women [27]. Vitamin B12 (cobalamin) has an important role in erythropoiesis and in conversion of homocysteine to methionine as cofactor of the methionine synthesis. Low B12 levels are related to hyperomocysteinemia and high homocysteine level is associated with NTDs [28].

The water-soluble vitamin C and the fat-soluble vitamin E are powerful antioxidants. It has been demonstrated that they may reduce the risk of NTDs [29] and lower vitamin C serum levels have been reported in women with previous pregnancies affected by NTDs [30]. The risk of NTDs seems also to decrease with increasing maternal preconception zinc intake [31]. Moreover, several studies demonstrated that maternal diet highest in thiamin [27], iron [27], niacin [27], methionine [32], betaine [27, 33], choline [33-34] can contribute to reduce the risk of NTDs.

Other maternal conditions may cause low folate levels. Alcoholism is associated with impaired folate status and chronic ethanol ingestion is often associated with folate deficiency, dietary inadequacy and intestinal malabsorption [35]. About celiac disease, the studies indicate the presence of low serum folic acid concentration in patients suffering from this condition, due to impairment in intestinal folic acid absorption [36].

Many studies have found the elevated concentration of homocysteine in plasma and colonic mucosa of patients suffering from inflammatory bowel disease [37] and low levels of circulating vitamin B6, B12, and folate in these patients [38]. These nutritional deficiencies could result from inadequate nutritional intake, malabsroption due to intestinal mucosal damage, greater nutritional requirements related to bowel inflammation and intestinal resection [39].

Pregestational Diabetes

Pregestational diabetes has been associated with several congenital malformation, but NTDs, (including spina bifida, anencephaly, encephalocele, holoprosencephaly), and cardiovascular defects, are among the most commonly observed [40-43].

A recent study presented data from 18 population-based EUROCAT registries of congenital anomalies in 1990-2005 and showed significantly increased ORs for anencephaly (OR= 1.57, 95% CI= 1.14-2.16) and encephalocele (OR=1.90, 95% CI=1.20-3.02), but not for spina bifida (OR=3.27, 95% CI= 1.67-6.39) [44].

In animal models, chemical induction of diabetes by streptozotocin, lead to NTDs in 10-15% of exposed mice [45-46]. High dietary glycemic index and dietary glycemic load seems both to be associated with an increased risk of NTDs [47].

Maternal hyperglycemia can induce congenital malformations throughout different mechanisms: causing alterations in the cells lipid metabolism; because of the excess generation of reactive oxygen species (ROS) due to mitochondrial dysfunction and for the activation of programmed cell death, or apoptotic, signaling cascades [48]. Membrane fatty acid alterations can contribute to diabetic embryopathy and diabetic pregnant rats fed with safflower oil supplementation, which contains an high concentration of Arachidonic Acid (a polyunsaturated fatty acid present in the phospholipids of membranes of the body's cells), had a significantly reduced incidence of NTDs in their offspring compared with non-treated diabetic controls (7.6% vs. 20.0%, respectively) [49]. Superoxide dismutase (SOD) catalyzes the dismutation of ROS into oxygen and hydrogen peroxide. It has been demonstrated that diabetic transgenic mice, bearing the human copper zinc superoxide dismutase gene, has a lower incidence of NTDs compared with nontransgenic diabetic controls [50]. Moreover, oxidative stress inhibits the expression of Pax3, a gene essential for neural tube closure, [51]. Caspases are a family of cysteine proteases that play essential roles in apoptosis, necrosis, and inflammation and high glucose levels have been shown to cause their activation. Caspase-8 was shown to be expressed in the developing neural tube and its inhibition in high glucose-challenged embryos reduced the rate of embryonic malformation, including NTDs [52]. Moreover, maternal diabetes can causes NTDs via activating protein kinase Cs (PKCs), which regulate apoptosis. Cao et al induced diabetes in mice by intravenous injection of streptozotocin and demonstrated that, compared to wild type, the PKCδ homozygous (*pkcδ*$^{-/-}$) embryos had significantly lower NTD rate [53].

Hendriks et al. studied a population of 149 case and 178 control Mexican American women founding that hyperinsulinemia was a risk factor for neural tube defects (OR = 1.91, 95% , CI = 1.21-3.01). Adjustment for obesity only slightly reduced the effect of hyperinsulinemia (OR = 1.75, 95% CI = 1.09-2.82) while a modest effect remained for obesity after adjustment for hyperinsulinemia (OR = 1.45, 95% CI = 0.84-2.51) [54].

Maternal Obesity

Obesity is currently a worldwide epidemic, and there are significant health implications of maternal obesity for both mother and child, including an increased risk of congenital malformation. Neural tube defects were the most frequently investigated congenital anomaly group. This association can be due to several causes:

1. Maternal obesity is often associated to pregestational diabetes, a well known risk factor of NTDs.
2. Nutritional/micronutrient deficiency, in particular folate deficiency can be observed in obese mothers.
3. Prenatal diagnosis of NTDs can be more complicated because of technical difficulty during fetal ultrasound screening and this can lead to a reduced rate of pregnancy termination among obese mothers [55].

In 1994, Waller et al. published the first work demonstrating that offspring of obese women are at an increased risk of neural tube defects and several other malformations [56]. In 2009 Stothard et al. conduced a systematic review and a meta-analysis of 39 studies to assess current evidence of the association between maternal overweight, maternal obesity, and congenital anomaly. Obese and Overweight mothers had a significantly increased risk of a pregnancy affected by a neural tube defect compared with normal-weighted women (OR respectively 1.87, 95% CI= 1.62-2.15; OR=1.20, 95% CI= 1.04-1.38), with a higher prevalence of both spina bifida and anencephaly [57]. Moreover, it has been demonstrated that a diet with ≥0,4 mg of folate per day is not protective against neural tube defects in women who weigh >70 kg [56, 58-59] and the risk of neural tube defects in the offspring of obese women has not been decreased by the widespread folic acid fortification of food in the 1990s [59-60]. An observational, study involving 129 pregnant women in 16 to 18 weeks gestation period found a significant association between dyslipidemia (especially hypertriglyceridemia) and NTDs. The OR for neural tube defects risk in dyslipidemic cases was 24.23 (95% CI= 4.73 -148.60) with a relative risk of 12.12 [61]. A case control study of 40,000 women showed that obese women were three times more likely to have a baby with a NTD [62]. Weight gain during pregnancy is usually lower among obese than nonobese women; low maternal weight gain during pregnancy (<10 kg) has also been associated with increased NTDs risk, although the association was similar among obese and nonobese women [63]. Carmichael et al. reported that the odds of a NTD-affected pregnancy was 30–50% lower among women who were physically active during the periconception period, and obese women are often less likely to do exercise [64]. The metabolic syndrome is variably defined as the concurrence of some combination of abdominal obesity, aberrant glycemic control, hyperlipidemia, and hypertension [65] and it is associated with oxidative stress, a well known risk factor for NTDs.

Smoke

Nicotine can interfere with embryogenesis increasing apoptosis throughout increased ROS e Ca^{2+} levels, in a concentration-dependent manner [66]. It has been demonstrated that smokers have lower folate concentration in plasma and decreased plasmatic level of some vitamins such as B12, vitamin B6, and antioxidant vitamins or micronutrients vs non-smokers [67-70]. Exposure to tobacco has been related with hyperomocysteinemia both in animals [71] and humans [72]. Literature data regarding possible association between NTDs and tobacco exposure during pregnancy are conflicting. Many studies have not demonstrated an increased risk of NTDs in the offspring of mothers who smoked cigarettes during pregnancy [73-75], although other studies found an association between NTDs and maternal smoking, with a reported OR which ranged from 1.3 to 1.8 [76-78]. A case control study based on National Birth Defects Prevention Study database compared mothers exposed only to passive smoke with nonsmokers (and also not exposed to passive cigarette smoke), founding that the first category of omen had an increased NTD risk (OR= 1.7, 95% CI= 1.4-2.0) but revealing no association between a moderate amount of maternal cigarette smoking (15–24 cigarettes per day) and the occurrence of NTDs in offspring. This risk was modest and non significant (OR= 1.6, 95% CI= 0.9-3.0) among mothers who smoked 25 or more cigarettes a day during the periconception period [79]. A Chinese study investigated the effect of smoke exposure from a partner on the risk of having a newborn baby with NTDs in non-smoking women, founding that the risk of NTDs was 1.4 (0.9 to 2.0), 1.8 (1.2 to 2.6), 1.9 (1.2 to 3.0) and 2.7 (1.6 to 4.7) for partner smoking of <1, 1-9, 10-19 and> 20 cigarette/ day, respectively [80].

Drugs

Epilepsy affects half a million women of child-bearing age worldwide [81]. Generalized tonic-clonic seizures can induce fetal lactic acidosis [82]; hypoxia and status epilepticus can cause fetal death [83]. Moreover, recurrent tonic-clonic seizures during pregnancy have been associated with poor cognitive performance in childhood [84-85]. Management of epilepsy in pregnant women is difficult because of the need of balancing these risks and to select a treatment that is effective in preventing major seizures while minimizing adverse fetal drug effects. The first report that Antiepileptic drugs (AEDs) may have teratogenic effects dates back over forty years. Large epidemiological studies have demonstrated an increase in major malformation of 4-7%, compared to the general population rate of 2-3% [86]. Polytherapy seems to be more harmful than single agent therapy, especially the combination of carbamazepine, valproic acid and Phenobarbital. Exposure to AEDs has been associated with congenital heart disease, cleft lip or palate, urogenital defects and NTDs [87-88].

Valproic acid (VPA) is efficient against several types of epilepsy and it is also used as a mood stabilizer. The use of VPA during pregnancy is associated with a 1–2% incidence of NTDs, mainly with lumbosacral meningomyelocele (spina bifida aperta). This association was already known in the 1980s and was confirmed in the following years by numerous case-control studies and cohorts [89-91]. It is also associated with congenital heart diseases, limb defects and facial abnormalities [92]. VPA acts specifically as an antimetabolite to folic acid and it could causes possible changes in gene expression in the embryo [93]. The congenital anomaly risk of VPA appears to be dose-related. Daily doses of VPA greater than 1000 mg or 1100 mg sodium valproate may significantly increase the risk of congenital malformations compared to lower dose [94].

Carbamazepine is a folic acid antagonist used for the treatment of epilepsy, chronic pain and mood disorders and it has also been related with congenital defects, in particular with NTDs (0.6–1.7%) [95]. The National Birth Defects Prevention Study found an increased risk for neural tube defects following first trimester exposure to carbamazepine, independent of periconception FA use, with an adjusted odds ratio of 5.0 (95% CI= 1.9-12.7) [96]. In women with no crises in the previous two years, a gradual discontinuity of therapy before conception may be taken into consideration. This is not applied to women with juvenile myoclonic epilepsy, due to the high risk of relapse [97]. Moreover, fetal risks may be reduced monitoring maternal plasma concentrations of AEDs, adjusting dose based on seizure control and avoiding withdrawal or changes of AEDs when pregnancy is already established.

Methotrexate inhibits dihydrofolic acid reductase and it is used in neoplastic diseases, psoriasis, rheumatoid arthritis and ectopic pregnancy. Fetal abnormalities after methotrexate exposure have been demonstrated in chicks [98], mice [99], rats [100], and rabbits [101]. "Clover-leaf" skull has been associated with methotrexate in utero exposure in humans, as well as Limb defects, absent ossification centers, central nervous system (CNS) abnormalities including anencephaly, hydrocephaly, and meningomyelocele [102]. In a review of case reports, it has been suggested that the maternal methotrexate dose that is necessary to produce developmental anomalies is probably above 10 mg per week, and a critical period for exposure was described as 6 through 8 weeks after conception, although not universally accepted [103].

Trimethoprim is a folic acid antagonist used as an antimicrobial agent commonly administered in combination with a sulfonamide. Based on its mechanism of action, it should potentially increases the risk of NTDs in exposed offspring. A case-control study found an adjusted odds ratios of NTDs related to exposure during the first or second months after the last menstrual period of 4.8 (95% CI= 1.5- 16.1) for trimethoprim, based on five exposed cases [104].

Periconception use of weight loss products has also been associated with NTDs. Bitsko et al reported an increased risk of anencephaly in women using any weight loss products (aOR= 2.6, 95% CI= 1.3-5.3) and an increased aOR with anencephaly (aOR= 2.8, 95% CI=1.0-7.3) in women using ephedra [105]. Several mechanisms may explain these findings: a possible direct teratogenic effect on the embryo during early development of these products; uncorrected dieting behavior; other unmeasured behavioral risk factors among the women who reported using the products. Moreover, weight-loss during early pregnancy and low weight gain during pregnancy in themselves have been related with NTDs [106-107].

Efavirenz is a nonnucleoside reverse transcriptase inhibitor given orally at the standard dosage of 600 mg once daily to treat human immunodeficiency virus infection. It is usually used in combination therapy. The drug is classified by the US Food and Drug Administration in pregnancy category C [108]. Congenital malformation have been observed in 3 (15%) of 20 pregnant cynomolgus monkeys, exposed at therapeutic doses similar to those used for humans. These abnormalities included anencephaly and anophthalmia in one fetus, microphthalmia in the second fetus, and cleft palate in the third fetus [109]. In a pregnancy registry maintained by a collaboration of manufacturers of antiretroviral drugs,the overall incidence of congenital defects after in utero exposure to efavirenz was 2.6% (95 % CI 1.5-4.2%). In retrospective reports including efavirenz, 4 births involved infants with myelomeningocele and two involved Dandy-Walker defects [110]. De Santis et al. have reported on three pregnancies that included first trimester exposures to efavirenz in combination with other drugs, reporting a normal infant, a spontaneous abortion at 7 weeks of gestational age and a case of sacralmyelomeningoceles with cerebral ventriculomegaly [111]. A second case of myelomeningocele in an infant with first-trimester exposure to efavirenz was reported in 2005 by Saitoh et al [112]. A recent metanalysis and a Comprehensive national surveillance study in the UK and Ireland did not find elevated risks of congenital deformities from first trimester efavirenz exposures [113-114]. In 2010 a U.S. groupy identified 5 efavirenz children with congenital anomalies (laryngiomalacia (N=1), meningomyelocele with Arnold-Chiari Malformation Type II (N=1), hypospadias (N=1), varus feet and hypertonicity of extremities (N=1), and cleft palate (N=1) [115]. Their 2012 report found that the only statistically significant association of an antiretroviral with risk of congenital anomalies was for efavirenz exposure during the first trimester (aOR 2.84, 95% CI, 1.13-7.16), reporting a case of spina bifida cystic with Arnold-Chiari malformation in a series of the 47 infants with first trimester exposure to efavirenz [116]. In

February 2012, the British HIV Association proposed that efavirenz is acceptable for use during pregnancy [117].

Hyperthermia

Hyperthermia has been associated with congenital malformations in humans. It can produces adverse effects through direct action on the embryo or by producing maternal toxicity; however, animal studies suggest a direct effect [118]. A cohort study found that maternal fever in the first trimester was associated with an increased risk of NTDs (OR=2.9, 95% CI= 1.5–5.7). Fever-reducing medications lead to a lower risk effect (OR=2.4, 95% CI= 1.0 –5.6) in women who assumed the drugs than those who did not (OR, 3.8; 95% CI, 1.4-10.9). First-trimester maternal exposures to hot tubs, saunas, or electric blankets were associated with an OR of 3.6 (95% CI= 1.1-15.9) [119].

Shaw et al. also demonstrated that a maternal fever or febrile illness episode in the first trimester was associated with an increased risk for having an NTD-affected pregnancy with an odds ratio (OR) of 1.91 (90% CI, 1.35-2.72) for fever and an OR of 2.02 (95% CI, 1.20-3.43) for febrile illness [120]. A meta-analysis evaluating 15 studies reported an overall odds ratio for neural tube defects associated with maternal hyperthermia of 1.92 (95% confidence interval = 1.61-2.29) [121].

Prolonged sauna bathing could raise the body temperature above 39 °C, and clinical reports over the last decade have raised the suspicion regarding the possible danger of sauna bathing to the developing embryo. In Hot-tub bathing heat is absorbed more rapidly and it would be a more efficient means of elevating body temperature than saunas.

Milunsky et al. conducted a large-scale prospective epidemiologic study using a cohort of 23,491 women, founding that hyperthermic exposure limited to hot tub, sauna, fever or electric blanket in the first trimester of pregnancy was associated with an increased risk for NTDs. The relative risks (RRs) were 2.8 (95% CI=1.2-6.5) for hot tub use, 1.8 (95% CI, 0.4–7.9) for sauna, 1.8 (95% CI, 0.8–4.1) for fever, and 1.2 (95% CI= 0.5-2.6) for electric blanket [122].

A recent study examined the relationship between hot tub use and 17 major birth defects using data from The National Birth Defects Prevention Study, demonstrating that anencephaly was significantly more likely to report any use of a hot tub in early pregnancy (OR=1.68, 95% CI=1.05-2.70). Mothers who reported using a hot tub for longer than 30 min during early pregnancy had an OR of 2.91 (95% CI=1.13-7.46) for anencephaly, However, two studies reported no elevation in risk [123-124].

Parental Occupational Exposure and other Chemical Agents

Several studies focused on environmental exposure and a possible correlation with NTDs. Toxic chemical substances can cross the placenta and may be found in cord blood. Animal models studied the possible association between various types of pesticides and NTDs [125-126].Women who reported using pesticides in their homes or yards seems to be more likely to have NTD-affected pregnancies. Brender et al. found a two-times increased risk (95% CI= 1.2-3.1) in self-reported pesticide exposures within 0.25 mile of cultivated fields in a population of Mexican American women [127]. Another study found increased risks for maternal agricultural occupations (OR = 5.6, 95% CI = 1.8-17.8 and for cleaning women (OR = 19, 95% CI = 0.9-3.9) [128]. Shaw et al. reported that California women exposed to pesticides had an increased risk of pregnancy affected by NTDs for professional applications (OR= 1.6; 95% CI= 1.1-2.5), but showed a not significant association among women applying the pesticides themselves (OR= 1.1, 95% CI= 0.8-1.7) [129]. A case-control study carried out to explore associations between spina bifida and occupational exposure of the mother showed an increased risk for women working in agricultural occupations (OR = 3.4, CI=1.3-9.0) although no differences were seen in pesticide or disinfectant exposure among case and control mothers in agricultural occupations [130].

Mancozeb is fungicide used in agriculture, especially in Norway mainly in potato cultivation. It induces malformations of the neural tube in animal models [131-132]. A Norwegian study found a moderate association between NDTs (131 cases, prevalence 12.8/10 000 births), potato cultivation (Prevalence ratio= PR= 1.6, 95% CI= 1.1-2.3) and paternal work of > 500 hours/year (PR= 1.6, 95% CI= 1.1-2.5) [133].

Polycyclic aromatic hydrocarbons (PAHs) are lipophilic compounds formed during the incomplete burning of organic substances. Human exposure is common, through inhalation of tobacco smoke and other sources of combustion. Some of the highest exposure levels are found in the workplace [134-135].

These compounds can cross the placenta [136] and have been found in cord blood [137]. Data analyzed from 1997 to 2002 deliveries in the National Birth Defects Prevention Study, a large population-based case-control study in the United States, showed a suggested a positive association between maternal occupational exposure to PAHs and risk of NTDs in offspring. The crude OR for PAH exposure was 1.43 (95% CI= 0.92-2.22) for any NTD and 1.71 (95% CI= 1.03-2.83) for spina bifida. However, after adjustment, the ORs for NTDs and spina bifida were attenuated and no longer significant [138].

Organic solvents are commonly used in occupational settings as cleaners, degreasers and reagents in varied industrial processes. These compounds produce oxidative stress (OS) to which early embryonic development is strongly susceptible [139-140].

Using data from the population based National Birth Defects Prevention Study, Desrosiers et al. examined the relation between maternal occupational exposure to organic solvents. The prevalence of exposure to any solvent among mothers of NTD cases was 13.1% (n=511). Exposure to chlorinated solvents was associated with increased odds of NTDs (OR=1.96, CI= 1.34-2.87), especially spina bifida (OR=2.26, CI= 1.44- 3.53) [141].

Health care workers may be occupationally exposed to known and suspected teratogens including chemical, biological and physical agents. Matte et al. analyzed parental occupational histories for 4,915 case babies with congenital defects, registered during the years 1968-1980 by the Metropolitan Atlanta Congenital Defects Program (MACDP) registry, and for 3,027 control babies born. They found that offspring of mothers employed in a nursing occupation during the periconception period had an increased risk of having anencephaly or spina bifida (RR = 2.00, 95% CI= 1.01-4.30) and also found associations between NTDs and potential exposure to anesthetic gases and to x-radiation, but each association was based on only three case babies of potentially exposed parents [142]. These interesting findings required further investigations.

Certain paternal occupations and related exposures have been suggested as possible risk factors for NTDs. The increased interest on paternally mediated adverse pregnancy outcome comes from both human and animal studies. Various main mechanisms of male reproductive toxicity have been proposed: non-genetic (due to the presence of a chemical compound or drug in semen), genetic (gene mutation and/or chromosomal abnormality) and epigenetic (effect on gene expression, genomic imprinting or DNA methylation) [143]. A case–control study found a statistically significant positive associations for paternal occupational exposure to agrochemicals (OR = 2.72, 95% CI=1.14,

6.51) and animals (OR = 2.56, 95% CI= 1.06- 6.16). The occupations leading to the significantly raised association were farmers, gardeners and butchers [144].

Brender and Suarez reported a weak association between cases of anencephaly located in Texas between 1981 and 1986 and paternal exposure to pesticides (OR = 1.28, 95% CI= 0.8- 2.1) [145].

Aguilar-Garduño at al. showed that fathers who were exposed to organic solvents during the period between 3 months before and one after the date of the last menstruation had an higher risk of having a child with anencephaly (adjusted OR= 2.08, 95% CI= 0.93- 4.60). This augmented risk was also found for exposure during the last 5 years (adjusted OR= 2.16, 95% CI= 1.08-5.32 [146].

A meta-analysis conducted to assess the risks of spontaneous abortions (SAs) and major malformations (MMs) after paternal exposure to organic solvents analyzed five studies (N= 180,242) that examined NTDs, which in general demonstrated an increased risk. Relationships from both cohort and case-control studies were statistically significant and quantitatively, were essentially the same (1.78 and 1.92, respectively). For spina bifida there was an increased risk but it was not statistically significant, while for anencephaly a significant relationship was found [147].

Five independent studies have attributed an increased risk of NTDs in offspring of painters to exposure to organic solvents, metals and epoxy. This risk was estimated to be from 2.5 to 4.9 [145,148-151]. Irgens also found a non-significant association with spina bifida (OR= 2.03, 95% CI= 0.99-3.75) [152]. Blatter et al. published data from three register-based case- referent studies conducted in Sweden, Spain, and Hungary. The highest OR for spina bifida was found for bricklayers in Spain (OR= 2.8, 95% CI= 1.4- 5.4). Slightly increased ORs were found for paternal agricultural occupations in Sweden and Hungary (OR Sweden= 1.6, 95% CI= 0.6-4.5; OR Hungary= 1.3, 95% CI= 0.9-2.0). An increased but not significant risk was observed for the category of Swedish printers, painters, and paper and plastic workers (OR= 2,6; 95% CI= 0.7- 10.4. Slightly increased risks were found in Hungary for other industrial workers (OR= 1.3, 95% CI= 0.9-1.8) and transport workers (OR= 1.3, 95% CI=0.9-1.8) [153].

A population-based, case–control study of fetuses and liveborn infants with NTDs showed elevated OR for NTDs for the categories: technical, sales and administrative, OR = 1.5 (95% CI= 1.0, 2.4); service, OR = 2.0 95% CI=1.2- 3.1); farming, forestry and fishing, OR = 2.1 (95% CI 1.3-3.3); operators, fabricators and labourers, OR = 1.8 (95% CI 1.2- 2.7]; and military,

OR = 1.9 [95% CI= 0.7- 5.0]. Moreover, positive associations have been observed for NTD among the offspring of men who are drivers or members of the UK Armed Forces [148, 151, 154].

Conclusion

Periconception FA supplementation significantly decreases the risk of NTDs. Several studies demonstrated that planning a pregnancy increases a women's likelihood of taking FA supplements and gives women the opportunity to participate in preconception counseling (PC) [155-156].

In a study of Australian primigravid women with a planned birth, only 34.5 % of women began taking FA before pregnancy [157]. In the Netherlands, 61 % of women with a planned pregnancy took FA; this result was obtained after a national campaign began in 1995 [158]. These results agree with several previous studies, including a study of postpartum women from Maryland [159], and a 1999 Canadian survey [160]. A recent Italian study showed that only the 43.4 % of women participating to a telephone survey took FA before conception. Women who received PC were significantly more informed and more prone to take FA compared to women who did not receive any medical counseling before pregnancy (OR = 7.2, 95 % CI=4.6–11.4) [161]. PC was defined by CDC as: ''a set of interventions that aim to identify and modify biomedical, behavioral, and social risks to a woman's health or pregnancy outcome through prevention and management, emphasizing those factors that must be acted on before conception or early in pregnancy to have a maximal impact''[162].

Various maternal conditions, high- risk behavior, exposure to drugs or toxins may significantly influence pregnancy outcome, increasing the risk of NTDs in the offspring. Preconception supplementation of 0,4 mg of FA is not sufficient for this selected population in order to prevent NTDs. Women at high-risk may benefit from daily dose of 4 mg/daily starting before conception up to the end of the first trimester. But despite that, other measures of prevention should be considered. Preconception risk assessment plays an important role, reducing adverse pregnancy outcomes. Multidisciplinary evaluation has a primary importance to detect high-risk women, in order to identify adequate strategies during the preconception period.

References

[1] Blencowe, H; Cousens, S; Modell, B et al. Folic acid to reduce neonatal mortality from neural tube disorders. *Int J Epidemiol* , 2010 39 Suppl. 1, 110–121.

[2] Lucock, M; Lucock, Z. Folic acid fortification: A double-edged sword. *Current Opinion in Clinical Nutrition and Metabolic Care,* 2009 12, 555–564.

[3] Lawrence, MA; Chai, W; Kara, R et al. Examination of selected national policies towards mandatory folic acid fortification. *Nutrition Reviews,* 2009 67(Suppl 1), S73–S78.

[4] Main, DM; Mennuti, MT. Neural tube defects: issues in prenatal diagnosis and counselling. *Obstet Gynecol,* 1986 67, 1–16.

[5] Centers for Disease Control and Prevention (CDC). Spina Bifida and Anencephaly Before and After Folic Acid Mandate --- United States, 1995--1996 and 1999—2000. *MMWR Morb Mortal Wkly Rep.,* 2004 53(17),362-365.

[6] Busby, A; Armstrong, B; Dolk, H et al. Preventing neural tube defects in Europe: A missed opportunity. *Reprod Toxicol,* 2005 20,393–402.

[7] Lumley, J; Watson, L; Watson, M; Bower, C. Periconceptional supplementation with folate and/or multivitamins for preventing neural tube defects. *Cochrane Database of Systematic Reviews,* 2001 4, CD001056.

[8] MRC Vitamin Study Research Group. Prevention of neural tube defects: results of the Medical Research Council vitamin study. *Lancet,* 1991 338,131-137.

[9] [No authors listed]. Recommendations for the use of folic acid to reduce the number of cases of spina bifida and other neural tube defects. *MMWR Recomm Rep,* 1992 41(RR-14),1-7.

[10] ACOG Committee on Practice Bulletins. ACOG practice bulletin. Clinical management guidelines for obstetriciangynecologists. Number 44, July 2003. *Obstet Gynecol,* 2003 102, 203-213.

[11] Aitken, DA; Crossley, JA; Spencer, K. Prenatal screening for neural tube defects and aneuploidy. In: Rimoin DL, Connor JM, Pyeritz RE, Korf BR, eds. Emery and Rimoin's Principles and Practice of Medical Genetics, 5th edition. Philadelphia: *Elsevier Churchill Livingstone,* 2007:636–678.

[12] Grosse, SD; Collins JS. Folic acid supplementation and neural tube defect recurrence prevention. *Birth Defects Res A Clin Mol Teratol,* 2007 79,737–742.

[13] Centers for Disease Control (CDC). Use of folic acid for prevention of spina bifida and other neural tube defects – 1983–1991. *MMWR Morb Mortal Wkly Rep,* 1991 40,513-516.

[14] Moore, LL; Bradlee, ML; Singer, MR et al. Folate intake and the risk of neural tube defects: an estimation of dose-response. *Epidemiology,* 2003 14, 200-205.

[15] Rinsky-Eng, J; Miller, L. Knowledge, use, and education regarding folic acid supplementation: continuation study of women in Colorado who had a pregnancy affected by a neural tube defect. *Teratology,* 2002 66, S29-S31.

[16] Felkner, M; Suarez, L; Hendricks, K et al.. Implementation and outcomes of recommended folic acid supplementation in Mexican-American women with prior neural tube defect-affected pregnancies. *Prev Med,* 2005 40,867-871.

[17] Dean, JH; Collins, JS; Clary, NJ et al. Preventing the recurrence of neural tube defects in South Carolina. *Proc Greenwood Genet Center,* 2007 26, 40-44.

[18] De Santis, M; Quattrocchi, T; Mappa, I; Spagnuolo, T; Licameli, A; Chiaradia, G; De Luca, C. Folic Acid Use in Planned Pregnancy: An Italian Survey. *Matern Child Health* J, 2012 Jun 13. [Epub ahead of print]

[19] Czeizel, AE; Dudás, I; Paput, L; Bánhidy, F. Prevention of neural-tube defects with periconceptional folic acid, methylfolate, or multivitamins? *Ann Nutr Metab* 2011, 58(4), 263-71.

[20] Rothenberg, SP; da Costa, MP; Sequeira, JM; Cracco, J; Roberts, JL; Weedon, J et al. Autoantibodies against folate receptors in women with a pregnancy complicated by a neuraltubedefect. *N Engl J Med,* 2004 350(2), 134-142.

[21] Pei, L; Liu, J; Zhang, Y; Zhu, H; Ren A. Association of reduced folate carrier gene polymorphism and maternal folic acid use with neural tube defects. *Am J Med Genet B Neuropsychiatr Genet,* 2009 150B(6):874-878.

[22] De Marco, P; Merello, E; Calevo, MG; Mascelli, S; Raso, A; Cama, A et al. Evaluation of a methylenetetrahydrofolate- dehydrogenase 1958G>A polymorphism for neural tube defect risk. *J Hum Genet,* 2006 51(2):98-103.

[23] Johnson, WG; Stenroos, ES; Spychala, JR; Chatkupt, S; Ming, SX; Buyske, S. New 19 bp deletion polymorphism in intron-1 of dihydrofolate reductase (DHFR): a risk factor for spina bifi da acting in mothers during pregnancy? *Am J Med Genet A*, 2004 124A(4), 339-345.

[24] Botto, LD; Yang, Q. 5,10-Methylenetetrahydrofolate reductase gene variants and congenital anomalies: a HuGE review. *Am J Epidemiol,* 2000 151(9), 862-877.

[25] Wilcken, B; Bamforth, F; Li, Z et al. Geographical and ethnic variation of the 677C/T allele of 5,10 meth ylenetetrahydrofolate reductase (MTHFR): findings from over 7,000 newborns from 16 areas world wide. *J Med Genet*, 2003 40: 619–625.

[26] De Marco, P; Calevo, MG; Moroni, A; Arata, L; Merello, E; Finnell, RH et al. Study of MTHFR and MS polymorphisms as risk factors for NTD in the Italian population. *J Hum Genet,* 2002 47(6), 319-324.

[27] Chandler, AL; Hobbs, CA; Mosley, BS; Berry, RJ; Canfield, MA; Qi, YP; Siega-Riz, AM; Shaw GM; National Birth Defects Prevention Study. Neural tube defects and maternal intake of micronutrients related to one-carbon metabolism or antioxidant activity. *Birth Defects Res A Clin Mol Teratol,* 2012 Aug 29. [Epub ahead of print].

[28] Ray, JG; Blom, HJ. Vitamin B12 insufficiency and the risk of fetal Neural Tube Defects. *Q J Med,* 2003 96, 289–295.

[29] Loeken MR. Free radicals and birth defects. *J Matern Fetal Neonatal Med,* 2004 15, 6–14.

[30] Smithells, RW; Sheppard, S; Schorah, CJ. Vitamin deficiencies and neural tube defects. *Arch Dis Child,* 1976 51, 944–950.

[31] Velie, EM; Block, G; Shaw, GM; Samuels, SJ; Schaffer, DM; Kulldorff, M. Maternal supplemental and dietary zinc intake and the occurrence of Neural Tube Defects in California. *Am J Epidemiol,* 1999 150, 605–616.

[32] Shaw, GM; Velie, EM; Schaffer, DM. Is dietary intake of methionine associated with a reduction in risk for neural tube defect-affected pregnancies? *Teratology,* 1997 56, 295-299.

[33] Shaw, GM; Carmichael, SL; Yang, W et al. Periconceptional dietary intake of choline and betaine and neural tube defects in offspring. *Am J Epidemiol,* 2004 160, 102-109.

[34] Shaw, GM; Finnell, RH; Blom, HJ et al.. Choline and risk of neural tube defects in a folate-fortified population. *Epidemiology,* 2009 20, 714-719.

[35] Wani, NA; Hamid, A; Kaur, J. Folate status in various pathophysiological conditions. *IUBMB Life,* 2008 60(12), 834-42.

[36] Calvani, MJr; Parisi, P; Guaitolini, C; Parisi, G; Paolone, G. Latent coeliac disease in a child with epilepsy, cerebral calcifications, drug-induced systemic lupus erythematosus and intestinal folic acid malabsorption associated with impairment of folic acid transport across the blood-brain barrier. *Eur J Pediatr,* 2001 160, 288-292.

[37] Mahmood, A; Needham, J; Prosser, J. Prevalence of hyperhomocysteinaemia, activated protein C resistance and prothrombin gene mutation in inflammatory bowel disease. Eur. J. Gastroenterol. *Hepatol,* 2005 17, 739-744.

[38] Morgenstern, I; Raijmakers, MT; Peters, WH. Homocysteine, cysteine, and glutathione in human colonic mucosa: elevated levels of homocysteine in patients with inflammatory bowel disease. *Dig Dis Sci,* 2003 48, 2083-2090.

[39] Peyrin-Biroulet, L; Rodriguez-Guéant, RM; Chamaillard, M; Desreumaux, P; Xia, B; Bronowicki, J P; Bigard, MA; Guéant, JL. Vascular and cellular stress in inflammatory bowel disease: revisiting the role of homocysteine. *Am J Gastroenterol,* 2007 102, 1108- 1115.

[40] Becerra, JE; Khoury, MJ; Cordero JF et al. Diabetes mellitus during pregnancy and the risks for the specific birth defects: a populationbased case-control study. *Pediatrics,* 1990 85,1-9.

[41] Kalter H. Case reports of malformations associated with maternal diabetes: history and critique. *Clin Genet,* 1993 43(4), 174-179.

[42] Cohen, MM Jr; Shiota, K. Teratogenesis of holoprosencephaly. *Am J Med Genet,* 2002 109(1), 1-15.

[43] Correa, A; Gilboa, SM; Besser, LM et al. Diabetes mellitus and birth defects. *Am J Obstet Gynecol* 2008, 199(3), 237.e1-237.e239.

[44] Garne, E; Loane, M; Dolk, H; Barisic, I; Addor, MC; Arriola, L et al. Spectrum of congenital anomalies in pregnancies with pregestational diabetes. *Birth Defects Res A Clin Mol Teratol,* 2012 94(3), 134-140.

[45] Phelan, SA; Ito, M; Loeken, MR et al. Neural tube defects in embryos of diabetic mice: role of the Pax-3 gene and apoptosis. *Diabetes,* 1997 46, 1189–1197.

[46] Pavlinkova, G; Salbaum, JM; Kappen, C et al. Wnt signaling in caudal dysgenesis and diabetic embryopathy. *Birth Defects Res A Clin Mol Teratol,* 2008 82, 710–719.

[47] Yazdy, MM; Liu, S; Mitchell, AA; Werler, MM. Maternal dietary glycemic intake and the risk of neural tube defects. *Am J Epidemiol,* 2010 171(4), 407-414.

[48] Reece, EA. Diabetes-induced birth defects: what do we know? What can we do? *Curr Diab Rep,* 2012 12(1), 24-32.

[49] Wiznitzer, A; Ayalon, N; Hershkovitz, R et al. Lipoic acid prevention of neural tube defects in offspring of rats with streptozocin-induced diabetes. *Am J Obstet Gynecol,* 1999 180,188-193.

[50] Hagay, ZJ; Weiss, Y; Zusman, I; Peled-Kamar, M; Reece, EA; Eriksson, UJ; Groner, Y. Prevention of diabetes-associated embryopathy by overexpression of the free radical scavenger copper zinc superoxide dismutase in transgenic mouse embryos. *Am J Obstet Gynecol,* 1995 173(4), 1036-1041.

[51] Zabihi, S; Loeken, MR. Understanding diabetic teratogenesis: where are we now and where are we going? *Birth Defects Res A Clin Mol Teratol,* 2010 88(10), 779-790.

[52] Zhao, Z; Yang, P; Eckert, RL; Reece, EA. Caspase-8: a key role in the pathogenesis of diabetic embryopathy. *Birth Defects Res B Dev Reprod Toxicol,* 2009 86(1), 72-77.

[53] Cao, Y; Zhao, Z; Eckert, RL; Reece, EA. The essential role of protein kinase Cδ in diabetes-induced neural tube defects. *J Matern Fetal Neonatal Med,* 2012 25(10), 2020-204.

[54] Hendricks, KA; Nuno, OM; Suarez, L; Larsen, R. Effects of hyperinsulinemia and obesity on risk of neural tube defects among Mexican Americans. *Epidemiology*, 2001 12(6):630-605.

[55] Vasudevan, C; Renfrew, M; McGuire, W. Fetal and perinatal consequences of maternal obesity. *Arch Dis Child Fetal Neonatal Ed,* 2011 96(5), F378-382.

[56] Waller, DK; Mills, JL; Simpson, JL; Cunningham, GC; Conley, MR; Lassman, MR; Rhoads, GG .Are obese women at higher risk for producing malformed offspring? *Am J Obstet Gynecol,* 1994 170(2), 541-548.

[57] Stothard, KJ; Tennant, PW; Bell, R; Rankin, J. Maternal overweight and obesity and the risk of congenital anomalies: a systematic review and meta-analysis. *JAMA,* 2009 301(6), 636-650.

[58] Rasmussen, SA; Chu, SY; Kim, SY et al. Maternal obesity and risk of neural tube defects: A metaanalysis. *Am J Obstet Gynecol,* 2008 198, 611-619.

[59] Ray, JG; Wyatt, PR; Vermeulen, MJ et al. Greater maternal weight and the ongoing risk of neural tube defects after folic acid flour fortification. *Obstet Gynecol,* 2005 105, 261-265.

[60] Wilson, RD; Johnson, JA; Wyatt, P et al. Preconceptional vitamin/folic acid supplementation: The use of folic acid in combination with a multivitamin supplement for the prevention of neural tube defects and other congenital anomalies. *J Obstet Gynaecol Can,* 2007 12, 1003-1026.

[61] Gupta, S; Arora, S; Trivedi, SS; Singh, R. Dyslipidemia in pregnancy may contribute to increased risk of neural tube defects -a pilot study in north Indian population. *Indian Journal of Clinical Biochemistry*, 2009 24 (2), 150-154.

[62] Watkins, ML; Rasmussen, SA; Honein, MA; Botto, LD; Moore, CA. Maternal obesity and risk for birth defects. *Pediatrics,* 2003 111(5 Part 2), 1152-1158.

[63] Shaw, GM; Todoroff, K; Carmichael, SL et al. Lowered weight gain during pregnancy and risk of neural tube defects among offspring. *Int J Epidemiol,* 2001 30,60-65.

[64] Carmichael; SL; Shaw, GM; Neri, E et al. Physical activity and risk of neural tube defects. Maternal Child Health, 2002 J6, 151-157.

[65] Ford, ES; Giles,WH. A comparison of the prevalence of the metabolic syndrome using two proposed definitions. *Diabetes Care,* 2003 26, 575-581.

[66] Zhao, Z; Reece EA. Nicotine-induced embryonic malformations mediated by apoptosis from increasing intracellular calcium and oxidative stress. *Birth Defects Res B Dev Reprod Toxicol,* 2005 74(5), 383-391.

[67] Piyathilake, CJ; Macaluso, M; Hine, RJ; Vinter, DW; Richards, EW; Krumdieck, CL. Cigarette smoking, intracellular vitamin deficiency, and occurrence of micronuclei in epithelial cells of the buccal mucosa. *Cancer Epidemiol Biomark Prev,* 1995 4,751-758.

[68] Piyathilake, CJ; Macaluso, M; Hine, RJ; Richards, EW; Krumdieck, CL. Local and systemic effects of cigarette smoking on folate and vitamin B-12. *Am J Clin Nutr,* 1994 60, 559-566. ;

[69] Vardavas, CI; Linardakis, MK; Hatzis, CM; Malliaraki, N; Saris, WH; Kafatos, AG. Smoking status in relation to serum folate and dietary vitamin intake. *Tob Induc Dis,* 2008 4, 8.

[70] McDonald, SD. Perkins, SL; Jodouin, CA, Walker MC. Folate levels in pregnant women who smoke: an important gene/environment interaction. *Am J Obstet Gynecol,* 2002 187, 620-625.

[71] Davis, JA; Brown, AT, Chen, H et al. Cigarette smoke increases intimal hyperplasia and homocysteine in a rat carotid endarterectomy. *J Surg Res,* 2004 121, 69-75.

[72] Kim, DB; Oh, YS; Yoo, KD et al. Passive smoking in never-smokers is associated with increased plasma homocysteine levels. *Int Heart J,* 2010 51 183-187.

[73] Grewal, J; Carmichael, SL; Ma, C et al. Maternal periconceptional smoking and alcohol consumption and risk for select congenital anomalies. *Birth Defects Res A Clin Mol Teratol,* 2008 82 519-526

[74] Kallen K. Maternal smoking, body mass index, and neural tube defects. *Am J Epidemiol,* 1998 147, 1103-1111.

[75] Wasserman, CR; Shaw, GM; O'Malley, CD et al. Parental cigarette smoking and risk for congenital anomalies of the heart, neural tube, or limb. *Teratology,* 1996 53, 261-267.

[76] Kelsey, JL; Dwyer, T; Holford, TR et al. Maternal smoking and congenital malformations: an epidemiological study. *J Epidemiol Commun Health,* 1978 32, 102-107.

[77] Evans, DR; Newcombe, RG; Campbell, H. Maternal smoking habits and congenital malformations: a population study. *BMJ,* 1979 2,171-173.

[78] McDonald, AD; Armstrong, BG; Sloan, M. Cigarette, alcohol, and coffee consumption and congenital defects. *Am J Public Health,* 1992 82, 91-93.

[79] Suarez, L; Ramadhani, T; Felkner, M; Canfield, MA; Brender, JD; Romitti, PA; Sun, L. Maternal smoking, passive tobacco smoke, and neural tube defects. *Birth Defects Res A Clin Mol Teratol,* 2011 91(1), 29-33.

[80] Li Z; Zhang, L; Ye, R; Liu, J; Pei, L; Zheng, X; Ren A. Partner cigarette smoking and risk of neural tube defects among infants of non-smoking women in northern China. *Tob Control,* 2012 Jun 15. [Epub ahead of print]

[81] Harden, CL; Meador, KJ; Pennell, PB et al American Academy of Neurology; American Epilepsy Society. Management issues for women with epilepsy - Focus on pregnancy (an evidence-based review): II. Teratogenesis and perinatal outcomes: Report of the Quality Standards Subcommittee and Therapeutics and Technology Subcommittee of the American Academy of Neurology and the American Epilepsy Society. *Epilepsia,* 2009 50(5), 1237-1246.

[82] EURAP Study Group. Seizure control and treatment in pregnancy: observations from the EURAP epilepsy pregnancy registry. *Neurology,* 2006 66, 354-360.

[83] Adab, N; Tudur, SC; Vinten, J; Williamson, P; Winterbottom, J. Common antiepileptic drugs in pregnancy in women with epilepsy. *Cochrane Database Syst Rev,* 2004 3, CD004848.

[84] Tomson, T; Perucca, E,; Battino D. Navigating toward fetal and maternal health: the challenge of treating epilepsy in pregnancy. *Epilepsia,* 2004 45,1171-75.

[85] Kaplan, PW; Norwitz, ER; Ben-Menachem, E et al. Obstetric risks for women with epilepsy during pregnancy. *Epilepsy Behav*, 2007 11, 283-91.

[86] Pennell, PB. Antiepileptic drugs during pregnancy: what is known and which AEDs seem to be safest? *Epilepsia,* 2008 49(suppl. 9), 43-55.

[87] Canger, R; Battino, D; Canevini, MP; Fumarola, C; Guidolin, L; Vignoli, A, et al. Malformations in offspring of women with epilepsy: a prospective study. *Epilepsia* 1999 40(9), 1231-1236.

[88] Arpino, C; Brescianini, S; Robert, EE; Castillo, E; Cocchi, G; Cornel, M, et al. Teratogenic effects of antiepileptic drugs: use of an international database on malformations and drug exposure (MADRE). *Epilepsia,* 2000 41(11), 1436-1443.

[89] DiLiberti, JH; Farndon, PA; Dennis, NR; Curry, CJ. The fetal valproate syndrome. *Am J Med Genet,* 1984 19, 473-481.

[90] Jager-Roman, E; Deichl, A; Jakob, S et al. Fetal growth, major malformations, and minor anomalies in infants born to women receiving valproic acid. *J Pediatr,* 1986 108, 997–1004.

[91] Ardinger, HH; Atkin, JF; Blackston, RD et al. Verification of the fetal valproate syndrome phenotype. *Am J Med Genet,* 1988 29, 171-185.

[92] Tatum, WO. Use of antiepileptic drugs in pregnancy. *Expert Rev Neurother,* 2006 6(7), 1077–1086.

[93] Ornoy A. Valproic acid in pregnancy: How much are we endangering the embryo and fetus? *Reproductive Toxicology,* 2009 28, 1-10.

[94] REPROTOX®. Valproic Acid. Available from http://www.reprotox.org/Members/AgentDetail.aspx?a=1318). Accessed October 2012.

[95] Pearse, SB, Rodriguez, LAG; Hartwell, C; Russell, G. A pregnancy register of patients receiving carbamazepine in the UK. *Pharmacoepidemiol Drug Saf,* 1992 1, 321-325.

[96] Werler, MM; Ahrens, KA; Bosco, JL; Mitchell, AA; Anderka, MT; Gilboa, SM et al. Use of antiepileptic medications in pregnancy in relation to risks of birth defects. *Ann Epidemiol,* 2011 21(11), 842-850.

[97] Thomas, SV. Management of epilepsy and pregnancy. *J Postgrad Med,* 2006 52(1), 57-64.

[98] Zamenhof, S. Differential effects of antifolate on the development of brain parts in chick embryos. *Growth,* 1985 49,28-33.

[99] Lanoue, L; Sulik, KK. Methotrexate (MTX) can induce neural tube defects (NTDs) in mouse embryos. *Teratology,* 1995 51,189.

[100] Schmid, BP. Monitoring of organ formation in rat embryos after in vitro exposure to azathioprine, mercaptopurine, methotrexate or cyclosporin A. *Toxicology,* 1984 31, 9-21.

[101] DeSesso, JM; Goeringer, GC. Methotrexate-induced developmental toxicity in rabbits is ameliorated by 1-(p-tosyl)-3,4,4-trimethylimidazolidine, a functional analog for tetrahydrofolate-mediated one-carbon transfer. *Teratology,* 1992 45, 271-283.

[102] REPROTOX®. Methotrexate. Available from http://reprotox.org/Members/AgentDetail.aspx?a=1036 Accessed October 2012.

[103] Feldkamp, M; Carey, JC. Clinical teratology counseling and consultation case report: low dose methotrexate exposure in the early weeks of pregnancy. *Teratology,* 1993 47, 533-539.

[104] Hernndez-Diaz, S; Werler, MM; Walker, AM; Mitchell, AA. Neural tube defects in relation to use of folic acid antagonists during pregnancy. *Am J Epidemiol,* 2001 153, 961-968.

[105] Bitsko, RH; Reefhuis, J; Louik, C; Werler, M; Feldkamp, ML; Waller, DK; Frias, J; Honein MA. National Birth Defects Prevention Study. Periconceptional use of weight loss products including ephedra and the association with birth defects. *Birth Defects Res A Clin Mol Teratol,* 2008 82(8), 553-562.

[106] Robert, E; Francannet, C; Shaw, G. Neural tube defects and maternal weight reduction in early pregnancy. *Reprod Toxicol,* 1995 9, 57- 59.

[107] Shaw, GM; Todoroff, K; Carmichael, SL et al. Lowered weight gain during pregnancy and risk of neural tube defects among offspring. *Int J Epidemiol,* 2001 30, 60-65.

[108] Nightingale, SL. From the Food and Drug Administration. JAMA, 1998 280, 1472.

[109] REPROTOX®. Efavirenz. Available from (http://reprotox.org/Members/AgentDetail.aspx?a=4157). Accessed October 2012.

[110] Antiretroviral Pregnancy Registry Steering Committee. Antiretroviral Pregnancy Registry International Interim Report for 1 January 1989 through 31 July 2011. Wilmington, NC: Registry Coordinating Center; 2011. Available from URL: www.APRegistry.com .

[111] De Santis, M; Carducci, B; De Santis, L; Cavaliere, AF; Straface G. Periconceptional exposure to efavirenz and neural tube defects. *Arch Intern Med,* 2002 162(3), 355.

[112] Saitoh, A; Hull, AD; Franklin, P; Spector, SA. Myelomeningocele in an infant with intrauterine exposure to efavirenz. *J Perinatol,* 2005 25, 555-556.

[113] Townsend, CL; Willey, BA; Cortina-Borja, M; Peckham, CS; Tookey, PA. *Antiretroviral therapy and congenital abnormalities in infants born to HIV-infected women in the UK and Ireland,* 1990-2007. AIDS. 2009 Feb 20;23(4):519-524.

[114] Ford, N; Calmy, A; Mofenson, L. Safety of efavirenz in the first trimester of pregnancy: an updated systematic review and meta-analysis. *AIDS,* 2011 25(18), 2301-2304.

[115] Brogly, SB; Abzug, MJ; Watts, DH; Cunningham, CK; Williams, PL; Oleske, J; Conway, D; Sperling, RS; Spiegel, H; Van Dyke, RB. Birth defects among children born to human immunodeficiency virus-infected women: pediatric AIDS clinical trials protocols 219 and 219C. *Pediatr Infect Dis J,* 2010 29(8), 721-727.

[116] Knapp, KM; Brogly, SB; Muenz, DG; Spiegel, HM; Conway, DH; Scott, GB; Talbot, JT; Shapiro, DE; Read JS. Prevalence of congenital anomalies in infants with in utero exposure to antiretrovirals. *Pediatr Infect Dis J,* 2012 31(2),164-170.

[117] (British HIV Association. Clinical Guidelines. Current Guidelines. Final version of guidelines (prior to publication). Management of HIV infection in pregnant women (2012). http://www.bhiva.org/PublishedandApproved.aspx. Accessed October 2012.

[118] Edwards MJ. Review: Hyperthermia and fever during pregnancy. *Birth Defects Res A Clin Mol Teratol,* 2006 76(7), 507-516.

[119] Suarez, L; Felkner, M; Hendricks, K. The effect of fever, febrile illnesses, and heat exposures on the risk of neural tube defects in a Texas-Mexico border population. *Birth Defects Res A Clin Mol Teratol,* 2004 70(10), 815-819.

[120] Shaw, GM; Todoroff, K; Velie, EM; Lammer EJ. Maternal illness, including fever and medication use as risk factors for neural tube defects. *Teratology,* 1998 57,1-7.

[121] Moretti, ME; Bar-Oz, B; Fried, S; Koren, G. Maternal hyperthermia and the risk for neural tube defects in offspring: systematic review and meta-analysis. *Epidemiology,* 2005 16(2), 216-219.

[122] Milunsky, A; Ulcickas, M; Rothman, KJ; Willett, W; Jick, SS; Jick, H. Maternal heat exposure and neural tube defects. *JAMA,* 1992 268,882-885.

[123] Chance, P. Hyperthermia and meningomyelocele and anencephaly. *Lancet,* 1978 1(8067),769-770

[124] Saxen, L; Holmberg, PC; Nurminen, M et al. Sauna and congenital defects. *Teratology,* 1982 25, 309-313.

[125] Chaineu, E; Binet, S; Pol, D; Chatellier, G; Meininger, V. Embryotoxic effectsof sodium arsenite and sodium arsenate on mouse embryos in culture. *Teratology,* 1990, 41,105-112.

[126] Gomes, J; Lloyd, OL; Hong, Z. Oral exposure of male and female mice to formulations of organophosphorus pesticides: congenital malformations. *Hum Exp Toxicol,* 2008 27, 231-240.

[127] Brender, JD; Felkner, M; Suarez, L; Canfield, MA; Henry JP. Maternal pesticide exposure and neural tube defects in Mexican Americans. *Ann Epidemiol,* 2010 20, 16-22.

[128] Blatter, BM; Roeleveld, N; Zielhuis, GA; Mullaart, RA; Gabreëls, FJ. Spina bifida and parental occupation. *Epidemiology,* 1996 7(2), 188-193.

[129] Shaw, GM; Wasserman, CR; O'Malley, CD; Nelson, V; Jackson RJ. Maternal pesticide exposure from multiple sources and selected congenital anomalies. *Epidemiology,* 1999 10, 60-66.

[130] Blatter, BM; Roeleveld, N; Zielhuis, GA; Gabreëls, FJ; Verbeek, AL. Maternal occupational exposure during pregnancy and the risk of spina bifida. *Occup Environ Med,* 1996 53(2), 80-86.

[131] Sato, K; Nakagata, N; Hung, CF; Wada, M; Shimoji, T; Ishii, S. Transplacental induction of myeloschisis associated with hindbrain crowding and other malformations in the central nervous system in Long-Evans rats. *Childs Nerv Syst,* 1985 1, 137-144.

[132] Ghate HV. Dithiocarbamate induced teratogenesis in frog embryo. *Riv Biol,* 1985 78, 288-91.

[133] Nordby, KC; Andersen, A; Irgens, LM; Kristensen, P. Indicators of mancozeb exposure in relation to thyroid cancer and neural tube defects in farmers' families. *Scand J Work Environ Health,* 2005 31(2), 89-96.

[134] Brandt, HC; Watson, WP. Monitoring human occupational and environmental exposures to polycyclic aromatic compounds. *Ann Occup Hyg,* 2003 47, 349-378.

[135] Hansen, AM; Mathiesen, L; Pedersen, M et al. Urinary 1-hydroxypyrene (1-HP) in environmental and occupational studies—a review. *Int J Hyg Environ Health,* 2008 211, 471-503.

[136] Gladen, BC; Zadorozhnaja, TD; Chislovska, N et al. Polycyclic aromatic hydrocarbons in placenta. *Hum Exp Toxicol,* 2000 19, 597-603.

[137] Madhavan, ND; Naidu, KA. Polycyclic aromatic hydrocarbons in placenta, maternal blood, umbilical cord blood and milk of Indian women. *Hum Exp Toxicol,* 1995 14, 503-506.

[138] Langlois, PH; Hoyt, AT; Lupo, PJ; Lawson, CC; Waters, MA; Desrosiers, TA; Shaw, GM; Romitti, PA; Lammer, EJ and the National Birth Defects Prevention Study. Maternal occupational exposure to polycyclic aromatic hydrocarbons and risk of neural tube defect-affected pregnancies. *Birth Defects Res A Clin Mol Teratol,* 2012 94(9), 693-700.

[139] Hansen, JM. Oxidative stress as a mechanism of teratogenesis. *Birth Defects Res C,* 2006 78, 293-307.

[140] Kovacic, P; Somanathan, R. Mechanism of teratogenesis: electron transfer, reactive oxygen species, and antioxidants. *Birth Defects Res* C, 2006 78, 308-325.

[141] Desrosiers, TA; Lawson, CC; Meyer, RE; Richardson, DB; Daniels, JL; Waters, MA; van Wijngaarden, E; Langlois, PH; Romitti, PA; Correa, A; Olshan, A. National Birth Defects Prevention Study. Maternal occupational exposure to organic solvents during early pregnancy and risks of neural tube defects and orofacial clefts. *Occup Environ Med,* 2012 69(7), 493-499.

[142] Matte, TD; Mulinare, J; Erickson, JD. Case-control study of congenital defects and parental employment in health care. *Am J Ind Med,* 1993 24(1), 11-23.

[143] De Santis, M; Cesari, E; Cavaliere, A; Ligato, MS; Nobili, E; Visconti, D; Caruso A. Paternal exposure and counselling: experience of a Teratology Information Service. *Reprod Toxicol,* 2008 26(1), 42-46.

[144] Fear, NT; Hey, K; Vincent, T; Murphy, M. Paternal occupation and neural tube defects: a case-control study based on the Oxford Record Linkage Study register. *Paediatr Perinat Epidemiol,* 2007 21(2), 163-168.

[145] Brender, JD; Suarez, L. Paternal occupation and anencephaly. *Am J Epidemiol,* 1990 131, 517-521.

[146] Aguilar-Garduño, C; Lacasaña, M; Blanco-Muñoz, J; Borja-Aburto, VH; García, AM. Parental occupational exposure to organic solvents and anencephaly in Mexico. *Occup Environ Med,* 2010 67(1), 32-37.

[147] Logman, JF; de Vries, LE; Hemels, ME; Khattak, S; Einarson TR. Paternal organic solvent exposure and adverse pregnancy outcomes: a meta-analysis. *Am J Ind Med,* 2005 47(1), 37-44.

[148] Fedrick, J. Anencephalus in the Oxford linkage study area. *Dev Med Child Neurol,* 1976 18, 643-656.

[149] Olshan, AF; Tschenke, K; Baird, PA. Paternal occupation and congenital anomalies in offspring. *Am J Ind Med,* 1991 20, 447-475.

[150] Olsen, J. Risk of exposure to teratogens amongst laboratory staff and painters. *Dan Med Bull,* 1982 30, 24-8.

[151] McDowall, ME. *Occupational reproductive epidemiology.* Studies on Medical and Population Subjects No. 50. London: Her Majesty's Stationery Office, 1985.

[152] Irgens, A; Krüger, K; Skorve, AH; Irgens, LM. Birth defects and paternal occupational exposure. Hypotheses tested in a record linkage based dataset. *Acta Obstet Gynecol Scand,* 2000 79(6), 465-470.

[153] Blatter, BM; Roeleveld, N; Bermejo, E; Martínez-Frías, ML; Siffel, C; Czeizel, AE. Spina bifida and parental occupation: results from three malformation monitoring programs in Europe. *Eur J Epidemiol,* 2000 16(4), 343-351.

[154] Office of Population Censuses and Surveys. *Congenital Malformations and Parents' Occupations.* OPCS Monitor MB3, 82/1. London, UK: HMSO, 1982.

[155] Arslan Ozkan, I; Mete, S. Pregnancy planning and antenatal health behaviour: Findings from one maternity unit in Turkey. *Midwifery,* 2010 26(3), 338-347.

[156] de Jong-van den Berg, LTW; Pharm, D; Hernandez-Diaz, S, Werler, MM et al. Trends and predictors of folic acid awareness and periconceptional use in pregnant women. *American Journal of Obstetrics and Gynecology,* 2005 192, 121-128.

[157] Wilton, D C; Foureur, MJ. A survey of folic acid use in primigravid women. *Women Birth,* 2010 23(2), 67-73.

[158] de Walle, HEK; de Jong-van den Berg, LTW. Ten years after the Dutch public health campaign on folic acid: The continuing challenge. *Eur J Clin Pharm,* 2008 64, 539-543.

[159] Chenga, D; Schwarzb, EB; Douglasb, E et al. Unintended pregnancy and associated maternal preconception, prenatal and postpartum behaviours. *Contraception,* 2009 79, 194-198.

[160] Neimanis, M; Paterson, J M; Bain, E. Preventing neural tube defects Survey of preconceptional use of folic acid. *Canadian Family Physician,* 1999 45, 1717-2216.

[161] De Santis, M; Quattrocchi, T; Mappa, I; Spagnuolo, T; Licameli, A; Chiaradia, G; De Luca, C. Folic Acid Use in Planned Pregnancy: An Italian Survey. *Matern Child Health J,* 2012 Jun 13. [Epub ahead of print].

[162] Posner, S F; Johnson, K; Parker, C et al. The National summit on preconception care: A summary of concepts and recommendations. *Maternal and Child Health Journal,* 2006 10, S197-S205.

In: Neural Tube Defects
Editor: Alberich Klein

ISBN: 978-1-62417-892-4

Chapter III

Understanding and Preventing Folic Acid-Resistant Neural Tube Defects

Pietro Cavalli[1]*, Vittorio Unfer[2], Sara De Grazia[2] and Giulio Cavalli[3]

[1]Medical Genetics, Azienda Istituti Ospitalieri, Cremona, Italy
[2]A.G.UN.CO., Obstetrics and Gynecology Center, Rome, Italy
[3]Internal Medicine and Clinical Immunology San Raffaele Hospital, Milano, Italy

Abstract

Neural tube defects (NTDs) are among the commonest birth defects, occurring in 0.5 to 2 per 1000 pregnancies. The neural tube is the embryonic structure that develops into the brain and spinal cord: the defects arise from an improper embryonic neural tube closure in the third and fourth weeks of pregnancy, causing malformations of the brain and spine (most commonly anencephaly and spina bifida).

Peri-conceptional supplementation with folic acid can prevent a proportion of NTD cases, however about 30% of NTDs are independent

* Corresponding author: Pietro Cavalli, MD, Medical Genetics, Azienda Istituti Ospitalieri, Via Concordia, 1. 26100 Cremona, Italy. Tel. +39 (0)372405217; fax +39 (0)372433999; E-mail: p.cavalli@ospedale.cremona.it.

from folate intake. Reports of NTD recurrence in families despite high dose of folate intake, further highlights the heterogeneity of NTD etiology, suggesting the existence of different NTD subtypes, many of them being folate-sensitive, and other folate-resistant. In those 'folate-resistant' cases no preventive therapy is available.

Experimental approach in mice revealed that NTDs in Pax3 and Cited2 mutant mice should be considered folate-sensitive, while NTDs in curly tail mutant mice are folate-resistant. Inositol is a vitamin required for the neural tube to close in mice, through activation of protein kinase C (PKC). Both D-chiro-inositol and myo-inositol prevent NTD pregnancies in curly tail mice, suggesting that its use should be effective in reducing experimental folate-resistant NTD recurrence.

Moreover, inositol supplementation significantly decreased the frequency of embryonic NTDs in diabetic rats. Inositol therapy appears safe in mouse pregnancy, without fetal defects or increased fetal loss.

Subsequently, lower inositol blood concentration was found in pregnant women carrying NTD fetuses, and inositol supplementation has been associated with healthy offspring in mothers at risk for folate-resistant NTDs.

Inositol therapy in humans has been used in treating psychiatric disorders, polycystic ovary syndrome without side effects. Children have been treated with inositol for autism and respiratory distress syndrome without side effects. High dosage of inositol was used: up to 18 g daily in adults, and 200 mg/kg in children.

Inositol is an isomer of glucose, is widespread in many foods, and is a constituent of living cells. It seems unlikely that exogenous inositol theraphy will pose a risk to the mother and/or the embryo/fetus.

Until now, 25 at high risk pregnancies have been successfully treated with myo-inositol and folic acid, without any side effects to the mother or the fetuses. More than 90 women were treated with up to 2gr inositol in the periconceptional period and reported no side effects.

All these evidences support the effectiveness and safety of periconceptional use of inositol associated with folic acid to further reduce the incidence of human NTDs.

Introduction

Neural tube defects (NTDs) are severe malformations of the CNS, occurring in 0.5 to 2 per 1,000 pregnancies. NTD results when the neural tube in the developing fetus fails to close in the third and fourth weeks of gestation. As pregnancy progresses, exposure to the amniotic fluid environment leads to neurodegeneration so that, at the time of birth, the exposed brain and/or spinal cord are severely damaged (Stiefel, 2007).

NTDs are usually classified as "closed" or "open", depending on the lesion being covered by skin or not. Cervical and thoracic regions are the least common sites, and lumbar and lumbosacral regions are the most common sites for those lesions. Open NTDs are mainly represented by spina bifida (myelomeningocele), that results from failure of closure in the spinal region, and anencephaly, the most severe form of NTDs, that is caused by a closure defect in the cranial segment of the neural tube. While anencephaly is a lethal condition, spina bifida is associated with physical and developmental disabilities, that can vary from mild to severe. Other open NTDs represent extremely rare conditions.

Closed or covered NTDs are represented by simple dysraphic conditions, as well as more complex states.

The etiology of NTDs is considered multifactorial in up to 70% of cases. Thus empiric risks are commonly used to predict the recurrence of NTDs (Main et al., 1986).

In other words, the square root of the prevalence of the disease can be used to calculate the recurrence risk in many multifactorial disorders.

After a pregnancy affected by NTDs, the recurrence risk is usually calculated in a figure of about 3-5%, the same as most multifactorial conditions (about 1/30) (Cowchock et al., 1980; Seller, 1981; Czeizel et al., 1984; Sebold et al., 2005).

After two affected pregnancies the recurrence risk is higher, and varies between 11.7 and 20% (1/9 – 1/5) (Seller, 1981; Czeizel et al., 1984).

More distant relatives, however, have a lower recurrence risk as their risk falls off very quickly with the degree of relationship.

NTDs can also be associated with chromosomal abnormalities, both aneuploidies and structural rearrangements. According to published data, 6.5 – 17% of fetuses affected by NTDs have chromosomal abnormalities (Babcook et al., 1995; Hume et al. 1996; Kennedy et al. 1998; Harmon et al., 1995; Hume et al., 1996; Coerdt et al., 1997; Kennedy et al.,1998).

Trisomy 18, trisomy 13, trisomy 9, X monosomy, trisomy 21, triploidy, unbalanced translocations, deletions, 22q11.2 microdeletion, have been associated with NTDs. Even isolated NTDs have been associated with chromosomal disorders (up to 16% of cases) (Harmon et al., 1995; Hume et al., 1996; Coerdt et al., 1997; Kennedy et al., 1998).

More recent molecular cytogenetic techniques (for example microarrays analysis) can detect more subtle chromosome abnormalities possibly associated with NTDs (Kirchhoff et al., 2009), and it is likely that the use microarray analysis and more sensitive molecular techniques will contribute to identify small chromosomal defects that have remained unidentified until now. Moreover familial clustering of NTDs has been observed, with probable autosomal recessive (AR), autosomal dominant (AD) and X-linked (XL) pattern of transmission (Lynch 2005; Rasmussen and Frias, 2005) In all these cases, no preventive therapy is effective, and a comprehensive medical approach is necessary to identify chromosomal abnormalities or genetic syndromes, and to accurately identify the etiology of NTDs.

Teratogenic insults, such as drug exposure during pregnancy, maternal metabolic conditions, and maternal infections, represent environmental factors that have been associated with an increased risk for NTD.

In particular, micronutrient deficiencies represent a broad spectrum of conditions associated with higher NTS risk: folate deficiency, leading to disturbance of folate metabolism; Inositol deficiency, leading to disturbance of phosphorylation events downstream of protein kinase C; Vitamin B12 deficiency, leading to disturbance of folate-related metabolism; Zinc deficiency, that interferes with folate absorbtion (Copp and Greene, 2010 (table 1).

All the conditions listed in table 1 should be considered while approaching a woman or a family with a previously affected pregnancy. In spite of the habit in modern medicine of treating individual patients according to an epidemiological approach, addressing at-risk individuals with a population-wide intervention, we believe that more effort is needed to target NTDs prevention by aetiological investigation and genetic counselling.

Table 1. Pathological conditions and teratogenic exposure causing low folate status and/or risk factors for NTDs

Metabolic conditions: maternal diabetes mellitus, obesity, hyperthermia
Drug exposure: anticonvulsivants (valproate, carbamazepine, phenobarbital, primidone, phenytoin), cocaine, vitamin A, isotretinoin, warfarin, thalidomide, methotrexate, diuretics, antihistamines, trimethoprim, sulfonamides, sulfasalazine, ergotamine, pyrimethamine, fumonisines, clomiphene(?), cigarette smoking, alcohol, caffeine, organic solvents, lead exposure
Maternal infections leading to fever/hyperthermia
Malabsorption: Inflammatory bowel disease, celiac disease, short bowel
Micronutrient deficiency: folate, inositol, vitamin B12, Zinc

It should be concluded that, while most of NTDs are universally considered multifactorial conditions, they might represent the tip of the iceberg of many different pathological conditions and diseases.

Thus, while counseling affected families or women, the etiology of every NTD must be investigated in depth, as NTDs could be associated with several genetic and non-genetic conditions, each requiring a different clinical approach.

Folic Acid and NTDs

Primary prevention with folic acid (FA) offers a possible universal solution to the problem of the morbidity and mortality caused by NTD.

Periconceptional folic acid supplementation can reduce both the occurrence and the recurrence risk for NTDs up to 70% (occurrence risk: OR 0.28 CI 95% 0.15-0.53; recurrence risk OR 0.68 (CI 95% 0.17-0.60) (De-Regil et al., 2010).

Folic acid fortification also reduces NTDs risk up to 50% (OR 0.54 CI 95% 0.46-0.63) (Blencowe et al., 2010).

Periconceptional folate supplementation is effective in preventing NTDs only when red blood cells folate (RBC) folate concentrations are greater than the level thought to be protective against NTDs (*i.e.*, ≥906 nmol/L).

Thus, before conception, RBC folate levels should be higher than 900 nmol/L, and blood folate should be higher than 15.9 nmol/L. (Bar-Oz et al. 2008; Daly et al. 1995).

The interactions between folic acid and zinc in the pathogenesis of NTDs are particularly intriguing. Zinc plays an essential role in the embryonic development, and maternal low levels of Zinc have been associated with an increased risk for fetal NTDs, as is the case of the genetic Zinc deficiency disease, acrodermatitis enteropathica (OMIM # 201100), which is associated with an increased risk for fetal malformation and NTDs is an example.

Zinc acts as a cofactor for the enzyme glutamate-hydroxilase that converts food polyglutamil folates into monoglutamyl folates, and its deficiency impairs intestinal absorbtion of folates.

Consistenly, folate deficiency may follow Zinc deficiency. Notoriously, a decrease in zinc blood concentration can be observed during pregnancy. So, the paradox of the very low blood folate levels in strict vegetarian women, despite their diet rich in folate and vitamins, might be explained by their chronic dietary Zinc deficiency.

Folate Supplementation

There is considerable variation in the estimates of the effect of FA supplementation on the prevalence of NTD. In the MRC trial, 1% of pregnancies still had recurrent NTDs, even after high dose therapy (4 mg daily) (1991).

The MRC NTD trial found that 4mg of folic acid administration in the periconceptional period prevents approximately 70% of NTDs. Further analysis based on a statistical model exploring the relationship between intake, background blood folate concentration and risk of NTDs, predicts that higher doses of folic acid may further reduce the risk of NTD (Wald et al., 2001).

It has been proposed that NTD risk is a function of the log(serum folate level) (Wald et al., 2001) thus suggesting that supplementation of folic acid with doses above the 4-5 mg (currently used for high risk pregnancies) could prevent even more than 70% of NTDs.

Wald and Coll. (Wald et al., 2001) found that, at the very same doses of folic acid supplementation, the reduction of NTD risk was greater in those women with the lower background folate concentration.

It is generally believed that the optimal physiological function of most nutrients depends on their concentration in blood and in tissues, as their activity normally increases with concentration. However, in individuals with an already optimal level of a given nutrient, an excessive intake may decrease the activity of the nutrient, and may exert toxic effects. Therefore those women with the lowest levels of folate are more prone to the beneficial effects of folic acid supplementation, while women with adequate folate blood levels are unlikely to experience any benefit from further folic acid supplementation. (Morris et al., 2011).

Even if periconceptional supplementation of folic acid at 0.4mg/day is widely considered effective in reducing NTD risk in the normal population, many women fail to achieve a protective folate blood concentration (Bar-Oz B et al., 2008), suggesting that other factors, for example polymorphisms in genes involved in folate pathway, should be taken into consideration before starting supplementation. Higher doses of folic acid should be given to women affected by diabetes, obesity, malabsorption disorders, or receiving treatment with antiepileptic drugs (Kennedy et al., 2012), although folic acid doses higher than 5 mg/day are usually not recommended.

As a result, and in a clinical setting, blood folate and homocysteine levels should be measured in every woman before pregnancy, and folate supplementation dosage should be tailored on individual baseline levels.

Mutations and polymorphisms in genes involved in folate pathway have been extensively investigated and in some studies have demonstrated an association with NTD risk. In some populations, the C677T MTHFR genotype has been associated with NTDs although these results have be proven inconsistent in different populations. (Blom et al., 2006).

Molecular analysis of the genes involved in the folate pathway was recommended by the US Preventive Services Task Force, as a way to evaluate the overall risk for NTDs (US preventive Services task Force, 2009). However that recommendation does not take into account the heterogeneity of NTDs, and folate gene analysis is probably not indicated in all cases of NTDs. Conversely, classification of NTD cases according to etiology and sensitivity or resistance to folate supplementation may result in a more effective patient-oriented clinical approach, while folate genotyping might be used to discriminate between folate-sensitive and folate-resistant NTDs.

Mothers homozygous for the C677T polymorphism in 5,10-methylene tetrahydrofolate reductase (MTHFR) gene have a 60% increased risk for an NTD-affected pregnancy (Kirke et al., 2004; Blom et al., 2006).

Therefore, genotyping for the C677T MTHFR allele should be considered in every woman with low serum folate levels. However, only women with biallelic C677T MTHFR mutations should benefit from higher periconceptional folate intake.

It must be stressed that not all cases of NTDs can be prevented by an increased intake of folic acid. As previously discussed, periconceptional administration of 4 mg folic acid prevents approximately 70% of NTDs (1991), thus suggesting that 30% of NTDs are not prevented by folic acid supplementation. This observation prompts the hypothesis that a proportion of human NTDs are inherently resistant to folic acid supplementation, and suggests a different classification of NTDs, based on the response to folate intake. "Folate-resistant" NTD cases may represent a different NTD subtype with different etiology from those cases that are prevented by folic acid. The report of multiple NTD cases in a single family despite high-dose folate intake is consistent with this 'folate-resistance' hypothesis. (Cavalli et al., 2002).

Similarly to non-multifactorial and syndromic NTDs whose causes are not linked to folate metabolism, isolated and multifactorial NTDs can recur in the same family, despite correct folic acid supplementation (Cavalli, 2002; Cavalli, 2008). In these cases, in accordance with the 'folate-resistance' hypothesis, the term "folate-resistant" NTD should be used.

Experimental studies confirm the existence of folate-resistant and folate-sensitive NTD subtypes. NTDs in mice homozygous for mutations of the Pax3,

Cart1, and crooked tail genes can be prevented with folic acid (Fleming and Copp, 1998) whereas NTDs in the curly tail mutant mouse are resistant to folic acid (Greene and Copp, 1997).

In an experimental model of folate-resistant NTD (the curly-tail mouse) both myo-inositol and D-chiro-inositol are effective in preventing NTD occurrence (Cogram et al., 2002), while folic acid is not effective.

Classification of NTDs

Anatomical classification groups NTDs into different categories: apart from "open" and "closed" (covered by skin) subtypes, NTDs might be further classified according to their anatomical aspect: Cranial, Caudal, Spina bifida, Anencephaly, Meningomyelocele, Encephalocele, Meningocele, Rachischisis, and other covered lesions.

A different classification, based on clinical findings, distinguishes between two main types: syndromic (associated with multiple congenital defects) and non-syndromic (isolated).

The latter classification seems to be useful in genetic counseling of affected families, as the recurrence risk is 3-5% in isolated cases, and up to 50% in syndromic conditions. However, while associated abnormalities can be easily identified after birth, the identification of associated abnormalities in the prenatal period may be difficult. In fact, according to the Eurocat Study about 91% of prenatal NTD cases are identified at a median gestation age of 17 weeks, and 84% of these affected pregnancies are terminated. It is worth mentioning that in the daily clinical practice, most of those pregnancies are not thoroughly investigated with ultrasound, and an accurate ultrasound diagnosis of the associated malformations is not always warranted (Boyd PA et al., 2008).

For this reason, that classifications seems unpractical and questionable from a clinical point of view, nor it seems useful in reproductive counselling.

A more convincing classification groups NTDs into two main types, depending on their response to folate intake, and is available in OMIM (Neural Tube Defects (MIM 182940) and Neural Tube Defects, folate-sensitive (MIM 601634) (Online Mendelian Inheritance in Man: http://www.ncbi.nlm.nih.gov/omim/).

This classification, based on the clinical response to folate periconceptional supplementation, highlights the functional relevance of folate deficiency. We believe that grouping NTDs according to their therapeutic and preventive approach may be useful in a clinical setting.

However, a different classification of NTDs, based on their etiology, should be pursued in daily medical practice, as it may allow a targeted prevention of individual NTD risk and may assist the clinician in the assessment of the recurrence risk. The following clinical aspects may be important for a classification based on NTDs etiology, and may assist the clinician in the assessment of the recurrence risk:

- Ultrasound fetal evaluation (associated abnormalities)
- Pre-postnatal fetal MRI
- Clinical/Genetic/Pathological evaluation (fetus/stillbirth/newborn)
- Detailed pregnancy history
- Accurate maternal medical history
- Medical family history (birth defects, fetal losses, stillbirth)
- Pedigree drawing (at least three generations); attention to parental consanguinity
- Cytogenetic analysis (fetus/stillbirth/newborn) (conventional/ molecular/microarrays)
- Parental cytogenetic analysis (conventional/molecular/microarrays)
- Maternal venous blood glucose concentration (oral glucose tolerance test) and HbA1c
- Maternal homocysteinemia, blood folate levels, red blood folate levels,
- Maternal folate enzymes genotyping (MTHFR, MTRR, CBS, MTHFD1, RFC1…) (single locus/multilocus analysis?)

To evaluate the first-occurrence risk in low-risk women, the following laboratory test should be required:

- Vitamin B12
- Vitamin B6
- homocysteinemia
- blood folate levels (red blood cells folate levels)
- Zinc
- anti-transglutaminase antibodies
- oral glucose tolerance test

Inositol and NTDs

Grainyhead-like-protein3 (*Grhl3)* knockout mice exhibit spina bifida with 100% penetrance, while milder phenotype is observed in ct/ct mice, suggesting that a reduced expression of *Grhl3* causes NTDs in the *curly tail (ct/ct)mouse model, and that ct is an hypomorphic allele of Grhl3).* Spina bifida in the ct/ct embryos results from the altered closure of the posterior neuropore, the posterior region of the neural tube in the caudal region of the embryo (Copp, 1985). Reduced and delayed closure are caused by an excessive ventral curvature of the posterior region of the embryo, which contains the neuropore (Brook et al., 1991). The balance between growth of anterior and posterior tissues regulates the angle of curvature. Hindgut endoderm in the *ct/ct* embryos shows an abnormally reduced cell proliferation rate, as well as in ventral tissues (Copp et al., 1988a). This under-proliferation leads directly to NTDs, as indicated by the finding that reducing the growth imbalance along the body axis can normalize curvature and decrease the frequency of spina bifida (Cogram et al., 2004). Neural tube closure can be normalized by inositol, that stimulates hindgut proliferation.

The relation between NTDs and inositol was first described in 1988 in rat embryos (Cockroft, 1988) and was confirmed in curly-tail mice a few years later (Seller, 1994). More recently, the hypothesis that inositol deficiency and/or the disruption of inositol signaling might be responsible for at least some subtypes of NTDs has been confirmed in several murine experimental models.

Inositol exerts a pivotal role in the cell biology, constituting the precursor for a number of second messengers. It was hypothesized that exogenous myo-inositol enters the inositol phospholipid cycle and is incorporated into inositol phospholipids, including phosphatidylinositol diphosphate (PIP2). External signals, via cell membrane receptors, induce the hydrolysis of PIP2, thus generating diacylglycerol (DAG), which activates protein kinase C (PKC) isoforms. The subsequent phosphorylation of specific substrates results in the downstream correction of that cell proliferation defect, which is responsible for spinal NTDs in *curly tail* embryos, thereby normalizing neural tube closure (Cogram et al., 2004).

Thus, PIP2 regulates the production of second messengers by its hydrolysis by phospholipase C and phosphatidylinositol 3-kinase, and this process is a key cellular event within all eukaryotic cells.

The preventive effect of inositol on NTD occurrence is not restricted to curly tail mice, since protection against diabetes-induced NTDs has been observed in other rodent models. Dietary inositol supplementation of 0.08 mg/kg/day is also effective in significantly decreasing the frequency of embryonic NTDs in diabetic rats (Reece et al.,1997). (Cogram et al.,2002). Noticeably, no evidence of an adverse effect of inositol either on pregnancy success or on fetal outcome was found in inositol-treated mice (Cogram et al., 2004).

Inositol 1,3,4-triphosphate5/6-kinase (ITPK1) plays a pivotal role in inositol metabolism, and represents a key regulatory enzyme for highly phosphorilated inositol phosphates including hexakisphosphate (IP6), which regulates intracellular regulation of ion channels, transcription, and DNA repair (Wilson et al., 1996; Wilson et al., 2009). ITPK1 is expressed in the brain, and is required for proper development of the neural tube. Disruption of inositol signaling might be responsible for NTDs in the genetically engineered mouse with a hypomorphic allele at ITPK1 (Majerus et al. 2010).

As a matter of fact, inositol 1,2,4-triphosphate 5/6-kinase (ITPK1) deficient mice develop NTDs, and hypomorphic mutations in ITPK1 gene are associated with 12% of mice embryos affected by NTDs.

Three isoforms of the enzyme. phosphatidylinositol-4-phosphate 5-kinase (PIP5KI) contribute to synthesize PIP2. It has been showed that disruption of PIP5KI interferes with normal developmental and cellular defects (Wang et al., 2007), and that PIP5KIγ-null embryos exhibit rostral neural tube closure defects (exencephaly) (Wang et al., 2007).

Moreover, studies performed to determine the expression of 1L-myo-inositol-1 phosphate synthase (MIP), leading to inositol-phosphate biosynthesis, reveal that inositol phosphate biosynthesis plays a key role in regulating the development of the mammalian brain (Alebous et al. 2009).

Hence, the animal data support a distinct inositol-dependent metabolic pathway that, when stimulated, can prevent the cellular dysfunction leading to spinal NTDs.

All together, these findings strongly suggest that inositol metabolism is essential for normal brain and CNS development and that disruption of inositol signaling is associated with NTDs (Wilson et al., 2009) (Majerus et al., 2010)

In humans, significantly lower inositol concentrations have been reported in the blood of mothers carrying NTD fetuses as compared with normal pregnancies and mothers with low blood levels of inositol show a 2.6-fold increased risk of carrying an affected offspring (Groenen et al., 2003).

Maternal diabetes is a well known risk factor for NTDs in offspring, and the relative risk for an affected NTDs pregnancy in diabetic women is 3-4 times higher than in non-diabetic women (Platt and Golde, 1983).

In those cases, NTDs is thought to originate from an impaired maternal metabolism leading to a defective turnover of phosphatidylinositol (Goto and Goldman, 1994).

Intriguingly, dietary inositol intake and NTD risk were not correlated statistically in a retrospective questionnaire analysis (Shaw et al., 2005). In a case of putative folate-resistant NTD, inositol supplementation was associated with a normal pregnancy outcome (Cavalli et al., 2002). These results suggest that inositol supplementation, but perhaps not normal dietary intake, could be beneficial in preventing folate-resistant NTDs in humans, as in rodents.

Other congenital defects were associated with inositol metabolism. As an example, non syndromic cleft palate has been also associated with inositol deficiency. Remarkably both NTDs and cleft palate (midline defects) are associated with neural crest cells migration defects (Krapels, 2004).

No official Recommended Daily Allowance (RDA) for myo-inositol, which is not recognized as a vitamin, has been reported. It is difficult to list recommended daily intakes as it is synthesized in the body: as a rough guide however, many nutritionists advise a daily consumption of 1,000 mg for adults.

Besides myo-inositol, as already mentioned, epimerization leads to the formation of up to nine stereoisomers, including d-chiro-inositol. There are a few dietary supplements that provide different amounts of these two isomers, despite this, the rationale of their dosage is not always so clear. The myo-inositol/d-chiro-inositol physiological ratio is specific in each tissue and the enzymatic conversion of the two isomers is affected by the action of an epimerase (Sun et al., 2002).

In blood, a ratio of 40:1 has been reported and, subsequently, the correct dosage of the two isomers in a dietary supplements should be based on this proportion (Nordio et al.,. 2012).

The use of inositol and folic acid in reducing NTD risk is based on a daily dose of 5mg folic acid and 1000mg myo-inositol, starting from 2 months before conception and continuing until 8 week s of pregnancy.

Up to now, 25 normal babies were born from high risk folate-resistant NTD pregnancies treated with inositol (Cavalli at al, 2011; Copp et al., 2011, Cavalli, unpublished), and follow-up is ongoing for 15 women at high risk for NTDs who became pregnant after FA and Inositol supplementation (Copp et al., 2011; Cavalli, unpublished).

Safety of Inositol

Inositol, chemically hexahydroxycylohexane, is any of nine stereoisomeric six-carbon sugar alcohols that closely resemble glucose in structure (cis-inositol, allo-inositol, epi-inositol, neo-inositol, myo-inositol, muco-inositol, scyllo-inositol, L-chiro-inositol, D-chiro-inositol).

Myo-inositol is the prevalent form of inositol in the body. It is available from various dietary sources, and can be synthesized from glucose by myo-inositol 1-phosphate synthase.

It is a constituent of many cell phosphoglycerides. Meso- or myoinositol, so named for its presence in muscle tissue, is biologically the most important isomer. Myo-Inositol is the precursor in the phosphatidylinositol cycle, a source of two second messengers (diacylglycerol and inositol triphosphate). Inositols and their phosphates lack a hydrolytically labile glycosidic linkage and are stable to degradative enzymes in vivo.

Inositol is widely found throughout mammalian tissues as phosphatidylinositol, and in cell membranes as phophoinositide (Hasan et al., 1974). It is classified as a member of the vitamin B complex (thiamine, riboflavin, niacin, pantothenic acid, biotin, pyridoxine, folic acid, inositol, and vitamin B12), although inositol is not a vitamin sensu stricto, as the human body can synthetize it. Inositol is an essential nutrient required by human cells in culture for growth and survival (Eagle et al., 1957). Moreover, its therapeutic use in humans has been extensively studied in autism, and other psychiatric disorders, in polycystic ovary syndrome, in patients on lithium therapy, and in respiratory distress syndrome (Benjamin et al., 1995; Colodny, 1998; Levine et al., 1997; Nestler et al., 1999; Howlett et al., 2003). In all these patients, no significant side effects of inositol therapy was reported, even at high doses (up to 18 g/day in psychiatric patients) (Palatnik et al., 2001). Based on more than 250 subjects who received treatment with, it has been concluded that the dosage up to 4g/day are completely free of side effects, whereas gastrointestinal side effects were associated with doses of 12g/day or higher (Carlomagno et al., 2011).

The safety of inositol usage in human pregnancies might be questioned, as inositol could mimic the effect of oxytocin, a nonapeptide hormone that induces uterine contractions at term. Oxytocin generates Ip3 via the activation of phospolipase C (PLC), thus inducing the release of calcium from the intracellular stores (Phaneuf et al., 1995). However, oxytocin effects further depend on modulation of calcium entry into the myometrial cells, a mechanism whose regulation is independent from IP3-mediated calcium release from sarcoplasmic reticulum (Sanborn et al., 1998).

A growing body of evidence suggests that administration of inositol during pregnancy is safe.

Inositol periconceptional supplementation was also tested as a way to induce ovulation in women affected by polycystic ovary syndrome (PCOS). In a group of 25 five PCOS affected women periconceptional administration of 4mg of myo-inositol was associated with a 40% pregnancy rate, and no increased in the incidence of abortion, nor collateral side affect or anomalous myometral contractions were reported (Papaleo et al. 2007).

Furthermore, 30 patients affected by polycystic ovary syndrome (PCOS) undergoing intracytoplasmic sperm injection (ICSI) received a daily dose of 4 gr myo-inositol in the periconceptional period, and in the eight clinical pregnancies that followed, no adverse effects or increase in the frequency of uterine contractions were mentioned. (Papaleo et al. 2009).

In one prospective clinical trial, 65 women undergoing in vitro fertilization (IVF) were randomized into two groups to receive 4 grams daily of myo-inositol plus folic acid or myo-inositol plus folic acid plus melatonin. A total of 22 pregnancies were obtained and no collateral effects imputable to inositol therapy were recorded (Rizzo et al. 2010).

In a recent, successfully concluded, randomized trial of diabetic pregnancies treated with 4 mg myo-inositol per day (listed in CliniclTrials.gov: Effect of Myo - Inositol Administration on Insulin Sensitivity in Gestational Diabetes Patients: NCT00734448. Available at http://clinicaltrials.gov/ct2/show/NCT00734448?term=inositolandrank=2) no collateral effects were reported by patients enrolled in the study (Corrado F. personal communication).

Myo-inositol administration was also tested in 54 women with polycystic ovary syndrome (PCOS), as a way to prevent gestational diabetes. Periconceptional supplementation with 4g/day of inositol was not associated with any side or collateral effects in those 46 women who became pregnant. (D'Anna et al., 2012).

Preliminary data from a double blind randomized clinical trial aimed at testing the hypothesis that the association of folic acid and inositol may prevent a greater proportion of NTDs than folic acid alone (The PONTI Trial: http://www.ich.ucl.ac.uk/ich/academicunits/Neural_development/Custom%20 Menu_03)), reported no collateral or side effect associated with periconceptional supplementation of 1g inositol per day (http://www.srhsb.org/nottingham-2011/the-ponti-clinical-trial-%E2%80%93-prevention-of-neural-tube-defects-by-inositol-%E2%80%93-an-update-on-progress.aspx). Among the 27 women were enrolled into the study, 15 women became pregnant and 5 babies were born without NTDs (Copp et al., 2011.)

In another study, seventeen pregnancies from 12 women at high risk for NTDs underwent periconceptional folic acid (5mg/day) and inositol (1g/day) supplementation starting at least four weeks before conception and seventeen babies unaffected by NTDs were born. No side effects were reported in that study (Cavalli at al., 2011).

In total, more than 90 women were treated with inositol (up to 4gr/day) in the periconceptional period, and no adverse effects was reported (Papaleo, 2007, 2009; Rizzo, 2010; Cavalli, 2011; Copp, 2011; D'Anna, 2012).

Collectively taken, these observations suggest that periconceptional inositol supplementation is unlikely to represent a risk for either the mother or the embryo/fetus.

In conclusion, the cumulative data from several studies of pregnant women during or after myo-inositol and folic acid treatment, strongly support the safety and effectiveness of periconceptional myo-inositol supplementation in reducing the overall risk for NTDs.

Inositol and Caffeine Intake

Maternal caffeine consumption has been associated with unfavorable effects on the fetus and the newborn. Other than fetal growth restriction (CARE study group, 2008) 2008), caffeine intake has also been considered as a risk factor for congenital malformations, particularly NTDs. (Schmidt et al., 2009, Schmidt et al., 2010).

Higher homocysteine blood levels will follow caffeine intake (Urgert et al., 2000; Verhoef et al., 2002) and hyperhomocysteinemia is associated with NTDs (Steegers-Theunissen et al.,1994), suggesting a possible explanation for the association between caffeine exposition and increased risk for NTDs.

However, given the inconsistencies often seen in the relationship between caffeine intake and the risk for NTDs (Linn et al., 1982; Rosenberg et al., 1982; Kurppa et al., 1983;) other possibile explanations involving different mechanisms need to be explored.

Caffeine can cross the placenta, and exposure to high doses of caffeine during pregnancy results in teratogenic effects, including NTDs, in animal studies (Wilkinson and Pollard, 1994; Marret et al., 1997).

Caffeine is rapidly absorbed and because of the absence of cytochrome P4501A2 (CYP1A2) in the placenta and the fetus, the amount of caffeine and metabolites available to the fetoplacental unit depends entirely on the maternal metabolism. (CARE study group, 2008; Schmidt et al., 2010). Genetic variability in metabolizing enzymes CYP1A2 and N-acetyltransferase 2 (NAT2) was proposed as an alternative explanation for the association between the increased risk for NTDs and caffeine intake (Schmidt et al., 2010). However caffeine inhibits IP3-related calcium channels in neural tissue (Bezprozvanny et al., 1994), thus suggesting the hypothesis of caffeine affecting a biological pathway involved in calcium regulation. (Cavalli et al., 2011)

Beside that specific mechanism, it is well known that caffeine may interfere with the bioavailability of several drugs (Benvenga et al. 2008).

The concomitant exposure to other dietary risk factors. for example low folate intake or alcohol consumption, results in an amplification of the teratogenic effects of caffeine (Ross and Persaud, 1989). That mechanism might be involved even in the complex interplay between caffeine, inositol, and the risk for NTDs.

In fact, NTDs etiology includes not only genetic and environmental factors, but even their joint effects. As both caffeine intake and inositol deficiency might increase the risk of occurrence and recurrence for NTDs in experimental models and in humans, their combined effect could be studied more in depth at the population level.

According to the reported positive effects of inositol in preventing NTDs in folate-resistant subjects, this interference could help clarifying the association between caffeine intake and NTDs development.

In fact, if caffeine may impair the absorption of inositol, the negative correlation between caffeine intake and the bioavailability of inositol, further strengthens the hypothesis of inositol deficiency as a risk factor for NTDs in humans.

Caffeine is likely to be one of the most used pharmacologic agent in the world, and even a small increase in risk for NTDs would have a great impact at the population level.

Therefore, to avoid caffeine interference during a prophylactic therapy for NTDs based on MI supplementation, women should be adviced about the effects of caffeine exposure in the periconceptional period and counseled to reduce their caffeine intake before conception. Morover, the concomitant intake of caffeine and inositol should be discouraged in women contemplating pregnancy and in their first months of pregnancy.

References

1991. Prevention of neural tube defects: results of the Medical Research Council Vitamin Study. MRC Vitamin Study Research Group. *Lancet* 338(8760):131-137.

Alebous, H. D., Cartee, R., Vaccari, D., Wright, O. A., Ahmed, A., Hood, R. D., Johnson, M. D. Developmental control of inositol phosphate biosynthesis is altered in the brain of both curly and phenotypically normal straight tail mutant mice. *Birth Defects Res. A Clin. Mol. Teratol.* 2009 Oct.;85(10):822-7

Babcook, C. J., Goldstein, R. B., Filly, R. A. Prenatally detected fetal myelomeningocele:is karyotype analysis warranted?. *Radiology* 1995; 194:491.4

Bar-Oz, B., Koren, G., Nguyen, P., Kapur, B. M. Folate fortification and supplementation-are we there yet? *Reprod. Toxicol.* 25 (2008) 408–412

Benjamin, J., Levine, J., Fux, M., Aviv, A., Levy, D., Belmaker, R. H. 1995. Double-blind, placebo controlled, crossover trial of inositol treatment for panic disorder. *Am. J. Psychiatry* 152(7):1084-1086.

Benvenga, S., Bartolone, L., Pappalardo, M. A., Russo, A., Lapa, D., Giorgianni, G., Saraceno, G., Trimarchi, F.Altered intestinal absorbtion of L-tiroxine caused by coffee. *Thyroid*. 2008 Mar.;18(3):293-301

Bezprozvanny, I., Bezprozvannaya, S., Ehrlich, B. E. Caffeine-induced inhibition of inositol(1,4,5)-triphosphate-gated calcium channels from cerebellum. *Mol. Biol. Cell.* 1994 Jan.;5(1):97-103.

Blencowe, H., Cousens, S., Modell, B., Lawn, J. Folic acid to reduce neonatal mortality from neural tube disorders. *Int. J. Epidemiol.* 2010; 39 (Suppl. 1):i110–i120.

Blom, H. J., Shaw, G. M., Den Heijer, M., and Finnell, R. H. (2006) Neural tube defects and folate: case far from closed. *Nat. Rev. Neurosci.*, 724-731).

Boyd, P. A., Devigan, C., Khoshnood, B., Loane, M., Garne, E., Dolk, H.; EUROCAT Working Group. Survey of prenatal screening policies in Europe for structural malformations and chromosome anaomalies, and their impact on detection and termination rates for neural tube defects and Down's syndrome. *BJOG*. 2008 May;115(6):689-96.

Brook, F. A., Shum, A. S., Van Straaten, H. W., Copp, A. J. Curvature of the caudal region is responsible fro failure of neural tube closure in the curly tail (ct) mouse embryo. *Development*. 1991 Oct.;113(2):671-8.

CARE study group. Maternal caffeine intake during pregnancy and risk of fetal growth restriction: a large prospective observational study. *BMJ.* 2008 Nov. 3;337:a2332

Carlomagno, et al., Inositol: safe clinical evidences, *European Review for Medical and Pharmacological Sciences*, 2011; 15: 931-936

Carlomagno, G., Unfer, V. Inositol safety: clinical evidences. *European Review for Medical and Pharmacological Sciences* 2011; 15: 931-936

Cavalli, P. and A. J. Copp, Inositol and folate resistant neural tube defects. *J. Med. Genet.,* 2002. 39(2): p. E5.

Cavalli, P., Cavallari, U., Unfer, V., Tonni, G. Caffeine intake and the risk of neural tube defects. *Birth Defects Res. A Clin. Mol. Teratol.* 2011 Jan.;91(1):67

Cavalli, P., S. Tedoldi and B. Riboli, Inositol supplementation in pregnancies at risk of apparently folate-resistant NTDs. *Birth Defects Res. A Clin. Mol. Teratol.,* 2008. 82(7): p. 540-2.

Cavalli, P., Tonni, G., Grosso, E., Poggiani, C. Effects of inositol supplementation in a color of mothers at risk of producing an NTD pregnancy. *Birth Defects Res. A Clin. Mol. Teratol.* 2011 91:962-5

Cockroft, D. L. Changes with gestational age in the nutritional requirements of postimplantation rat embryos in culture. *Teratology* (1988) 45:223-232

Coerdt, W., Miller, K., Holzgreve, W., Rauskolb, R., Schwinger, E., Rehder, H. Neural tube defects in chromosomally normal and abnormal human embryos. *Ultrasound Obstet. Gynecol.* 1997;10:410-5

Cogram, P., Hynes, A., Dunlevy, L. P., Greene, N. D., Copp, A. J. Specific isoforms of proteine kinase C are essential for prevention of folate-resistant neural tube defects by inositol. *Hum. Mol. Genet.* 2004 Jan. 1;13(1):7-14

Cogram, P., Tesh, S., Tesh, J., Wade, A., Allan, G., Greene, N. D., Copp, A. J. 2002. D-chiro-inositol is more effective than myo-inositol in preventing folate-resistant mouse neural tube defects. *Hum. Reprod.* 17(9):2451-2458.

Cogram, P., Tesh, S., Tesh, J., Wade, A., Allan, G., Greene, N. D., Copp, A. J. 2002. D-chiro-inositol is more effective than myo-inositol in preventing folate-resistant mouse neural tube defects. *Hum. Reprod.* 17(9):2451-2458.

Colodny, L., Hoffman, R. L. 1998. Inositol--clinical applications for exogenous use. *Altern. Med. Rev.* 3(6):432-447.

Copp, A. J. Relationship between timing of posterior neuropore closure and development of spinal neural tube defects in mutant (curly tail) and normal mouse embryos in culture. *J. Embryol. Exp. Morphol.* 1985 Aug.;88:39-54

Copp, A. J., Crolla, J. A., Brook, F. A. Prevention of spinal neural tube defects in the mouse embryo by growth retardation during neurulation. *Development.* 1988 Oct.;104(2):297-303.

Copp, A. J., Greene, D. E. Genetics and development of neural tube defects. *J. Pathol.* 2010; 220: 217-230)

Copp, A. J., Shepherd, V., Chitty, L., Greene, N. The PONTI clinical trial – Prevention of Neural Tube Defects by Inositol – an update on progress. *Presented at SRHSB, 55th annual meeting, Nottingham*, 2011

Cowchock, S., Ainbender, E., Prescott, G., Crandall, B., Lau, L., Heller, R., Muir, W. A., Kloza, E., Feigelson, M., Mennuti, M., Cederquist, L. The recurrence risk for neural tube defects in the United States: a collaborative study. *Am. J. Med. Genet.* 1980;5(3):309-14

Czeizel, A., Métneki, J. Recurrence risk after neural tube defects in a genetic counselling clinic. *J. Med. Genet.* 1984 Dec;21(6):413-6.

Daly, L. E., Kirke, P. N., Molloy, A., Weir, D. G., Scott, J. M. Folate levels and neural tube defects. Implications for prevention. *JAMA.* 1995;274:1698-1702

D'Anna, R., Di Benedetto, V., Rizzo, P., Raffone, E., Interdonato, M. L., Corrado, F., Di Benedetto, A. Myo-inositol may prevent gestational diabetes in PCOS women. *Gynecol. Endocrinol.*; 2012 Jun.;28(6):440-2.

De-Regil, L. M., Fernandez-Gaxiola, A. C., Dowswell, T., Pena-Rosas, J. P. Effects and safety of periconceptional folate supplementation for preventing birth defects. *Cochrane Database Sys Reviews.* 2010;10:CD007950

Eagle, H., Oyama, V. I., Levy, M., Freeman, A. E. Myo-inositol as an essential growth factor for normal and malignant cells in tissue culture. *J. Biol. Chem.* 1957 May;226(1):191-205

Fleming, A. and A. J. Copp, Embryonic folate metabolism and mouse neural tube defects. *Science*, 1998. 280(5372): p. 2107-9.

Goto, M. P., Goldman, A. S. Diabetic embryopathy. *Curr. Opin. Pediatr.* 1994 Aug.;6(4):486-91

Greene, N. D. and A. J. Copp, Inositol prevents folate-resistant neural tube defects in the mouse. *Nat. Med.*, 1997. 3(1): p. 60-6.

Groenen, P. M., Peer, P. G., Wevers, R. A., Swinkels, D. W., Franke, B., Mariman, E. C., Steegers Theunissen, R. P. 2003. Maternal myo-inositol, glucose, and zinc status is associated with the risk of offspring with spina bifida. *Am. J. Obstet. Gynecol.* 189(6):1713-1719.

Harmon, J. P., Hiett, A. K., Palmer, C. G., Golichowsli, A. M. Prenatal ultrasound detection of isolated neural tube defects: is cytogenetic evaluation warranted?. *Obstet. Gynecol.* 1995; 86:595-9

Hasan, S. H., Nishigaki, I., Tsutsui, Y., Yagi, K. Studies on myoinositol. IX. Morphological examination of the effect of massive doses of myoinositolon the liver and kidney of rat. *J. Nutr. Sci. Vitaminol.* (Tokyo). 1974 Feb.;20(1):55-8

Heseker, H. B., Mason, J. B., Selhub, J., Rosenberg, I. H., Jacques, P. F. Not all cases of neural tube defects can be prevented by increasing the intake of folic acid. *Br. J. Nutr.* 2009 Jul.;102(2):173-80

Howlett, A., Ohlsson, A. 2003. Inositol for respiratory distress syndrome in preterm infants. *Cochrane Database Syst. Rev.* (4):CD000366.

Hume, R. F., Drugan, A., Reichler, A., Lampinen, J., Martin, L. S., Johnson, M. P., Evans, M. I. Aneuploidy among prenatally detected neural tube defects. *Am. J. Med. Genet.* 1996; 61:171-3

Kennedy, D., Chitayat, D., Winsor, E. J., Silver, M., Toy, A. Prenatally diagnosed neural tube defects: ultrasound, chromosome, and autopsy or postnatal findings in 212 cases. *Am. J. Med. Genet.* 1998;77:317-21

Kennedy, D., Koren, G. Identifying women who may benefit from higher doses of folic acid in pregnancy. *Can. Fam. Physician.* 2012 Apr.; 58(4):394-7

Kirchhoff, M., Bisgaard, A. M., Stoeva, R., Dimitrov, B., Gillessen-Kaesbach, G., Fryns, J. P., Rose, H., Grozdanova, L., Ivanov, I., Keymolen, K., Fagerberg, C., Tranebjaerg, L., Skovby, F., Stefanova, M. Phenotype and 244k array-CGH characterization of chromosome 13q deletions: An update of the phenotypic map of 13q21.1-qter. *Am. J. Med. Genet.* A. 2009 149A(5):894-905.

Kirke, P. N., et al., Impact of the MTHFR C677T polymorphism on risk of neural tube defects: case-control study. *Bmj,* 2004. 328(7455): p. 1535-6.

Krapels, I. P., Rooij, I. A., Wevers, R. A., Zielhuis, G. A., Spauwen, P. H., Brussel, W., Steegers-Theunissen, R. P. Myo-inositol, glucose and zinc status as risk factors for non-syndromic cleft lip with or without cleft palate in offspring: a case-control study. *BJOG.* 2004 Jul.;111(7):661-8

Kurppa, K., Holmberg, P. C., Kuosma, E., Saxén, L. Coffee consumption during pregnancy and selected congenital malformations: a nationwide case-control study. *Am. J. Public Health.* 1983 Dec.;73(12):1397-9.

Larner, D-Chiro-Inositol – Its Functional Role in Insulin Action and Its Deficit in Insulin Resistance, *Int. Jnl. Experimental Diab. Res.*, 3:47-60, 2002

Levine, J., Aviram, A., Holan, A., Ring, A., Barak, Y., Belmaker, R. H. 1997. Inositol treatment of autism. *J. Neural Transm.* 104(2-3):307-310.

Lynch, S. A. Non multifactorial neural tube defects. *Am. J. Med. Genet.* Part C. 2005; 135C:69-76

Main, D. M., Mennuti, M. T. Neural Tube Defects: issues in prenatal diagnosis and counseling. *Obstet. Gynecol.* 1986 Jan.;67(1):1-16

Majerus, P. W., Wilson, D. B., Zhang, C., Nicholas, P. J., Wilson, M. P. 2010. Expression of inositol 1,3,4-trisphosphate 5/6-kinase (ITPK1) and its role in neural tube defects. *Adv. Enzyme Regul.* 50(1):365-372.

Marret, S., Gressens, P., Van-Maele-Fabry, G., Picard, J., Evrard, P. Caffeine-induced disturbances of early neurogenesis in whole mouse embryo cultures. *Brain Res.* 1997 Oct. 31;773(1-2):213-6.

Morris, M. C., Tangney, C. C. A potential design flaw of randomized trials of vitamin supplements. *Jama* 2011;305: 1348

Nestler, J. E., Jakubowicz, D. J., Reamer, P., Gunn, R. D., Allan, G. 1999. Ovulatory and metabolic effects of D-chiro-inositol in the polycystic ovary syndrome. *N Engl. J. Med.* 340(17):1314 1320.

Nordio M, Proietti E. The combined therapy with myo-inositol and D-chiro-inositol reduces the risk of metabolic disease in PCOS overweight patients compared to myo-inositol supplementation alone. Eur Rev Med Pharmacol Sci. 2012 May;16(5):575-81.

Palatnik, A., Frolov, K., Fux, M., Benjamin, J. 2001. Double-blind, controlled, crossover trial of inositol versus fluvoxamine for the treatment of panic disorder. *J. Clin. Psychopharmacol.* 21(3):335-339.

Papaleo, E., Unfer, V., Baillargeon, J. P., De Santis, L., Fusi, F., Brigante, C., Marelli, G., Cino, I., Redaelli, A., Ferrari, A. 2007. Myo-inositol in patients with polycystic ovary syndrome: a novel method for ovulation induction. *Gynecol. Endocrinol.* 23(12):700-703.

Papaleo, E., Unfer, V., Baillargeon, J. P., Fusi, F., Occhi, F., De Santis, L. 2009. Myo-inositol may improve oocyte quality in intracytoplasmic sperm injection cycles. A prospective, controlled, randomized trial. *Fertil. Steril.* 91(5):1750-1754.

Phaneuf, S., Europe-Finner, G. N., Carrasco, M. P., Hamilton, C. H., Lopez Bernal, A. 1995. Oxytocin signalling in human myometrium. *Adv. Exp. Med. Biol.* 395:453-467.

Platt, L. D., Golde, S. H. Neural tube defects in infants of diabetic mothers. *Am. J. Obstet. Gynecol.* 1983 Feb. 1;145(3):384-6

Rasmussen, S. A., Frias, J. L. Genetics of syndromic neural tube defects. In: Neural Tube Defects: from origin to treatment. *Wyzynski DF* ed. 2005 Oxford University Press, New York.

Reece, E. A., Khandelwal, M., Wu, Y. K., Borenstein, M. 1997. Dietary intake of myo-inositol and Neural tube defects in offspring of diabetic rats. *Am. J. Obstet. Gynecol.* 176(3):536-539.

Rizzo, P., Raffone, E., Benedetto, V. 2010. Effect of the treatment with myo-inositol plus folic acid plus melatonin in comparison with a treatment with myo-inositol plus folic acid on oocyte quality and pregnancy outcome in IVF cycles. A prospective, clinical trial. *Eur. Rev. Med. Pharmacol. Sci.* 14(6):555-561.

Rosenberg, L., Mitchell, A. A., Shapiro, S., Slone, D. Selected birth defects in relation to caffeine-containing beverages *JAMA*. 1982 Mar. 12;247(10):1429-32.

Ross, C. P., Persaud, T. V. Neural tube defects in early rat embryos following maternal treatment with ethanol and caffeine. *Anat. Anz.* 1989;169(4):247-52

Sanborn, B. M., Dodge, K., Monga, M., Qian, A., Wang, W., Yue, C. 1998. Molecular mechanisms regulating the effects of oxytocin on myometrial intracellular calcium. *Advances in experimental medicine and biology* 449:277-286.

Schmidt, R. J., Romitti, P. A., Burns, T. L., Browne, M. L., Druschel, C. M., Olney, R. S.; National Birth Defects Prevention Study. Maternal caffeine consumption and risk of neural tube defects. *Birth Defects Res. A Clin. Mol. Teratol.* 2009 Nov.;85(11):879-89.

Schmidt, R. J., Romitti, P. A., Burns, T. L., Murray, J. C., Browne, M. L., Druschel, C. M., Olney, R. S.; National Birth Defects Prevention Study. Caffeine, selected metabolic gene variants, and risk for neural tube defects. *Birth Defects Res. A Clin. Mol. Teratol.* 2010 Jul.;88(7):560-9.

Sebold, C. D., Melvin, E. C., Siegel, D., Mehltretter, L., Enterline, D. S., Nye, J. S., Kessler, J., Bassuk, A., Speer, M. C., George, T. M.; NTD Collaborative Group. Recurrence risks for neural tube defects in sibling of patients with lipomyelomeningocele. *Genet. Med.* 2005 Jan.;7(1):64-7.

Seller, M. J. Vitamins, folic acid and the cause and prevention of neural tube defects. *Ciba Found. Symp.* (1994);181:161-173

Seller, M. J., Recurrence risks for neural tube defects in a genetic counseling clinic population. *J. Med. Genet.*, 1981. 18(4): p. 245-8.

Shaw, G. M., Carmichael, S. L., Yang, W., Schaffer, D. M. 2005. Periconceptional dietary intake of myo-inositol and neural tube defects in offspring. *Birth Defects Res. A Clin. Mol. Teratol.* 73(3):184-187.

Steegers-Theunissen, R. P., Boers, G. H., Trijbels, F. J., Finkelstein, J. D., Blom, H. J., Thomas, C. M., Borm, G. F., Wouters, M. G., Eskes, T. K. Maternal hyperhomocysteinemia: a risk factor for neural tube defects? *Metabolism.* 1994 Dec.;43(12):1475-80.

Stiefel, D., Copp, A. J., Meuli, M. Fetal spina bifida in a mouse model: loss of neural function in utero. *J. Neurosurg.* (2007); 106:213-21)

Sun TH, Heimark DB, Nguygen T, Nadler JL, Larner J. Both myo-inositol to chiro-inositol epimerase activities and chiro-inositol to myo-inositol ratios are decreased in tissues of GK type 2 diabetic rats compared to Wistar controls. Biochem Biophys Res Commun. 2002 May 10;293(3):1092-8.

Urgert, R., van Vliet, T., Zock, P. L., Katan, M. B. Heavy coffee consumption and plasma homocysteine: a randomized controlled trial in healthy volunteers. *Am. J. Clin. Nutr.* 2000 Nov.;72(5):1107-10.

US Preventive Services Task Force. Folic acid for the prevention of neural Tube Defects: US preventive Services Task Force Recommendation Statement. *Ann. Int. Med.*, 2009; 150; p.626-631

Verhoef, P., Pasman, W. J., Van Vliet, T., Urgert, R., Katan, M. B. Contribution of caffeine to the homocysteine-raising effect of coffee: a randomized controlled trial in humans. *Am. J. Clin. Nutr.* 2002 Dec.;76(6):1244-8.

Wald, N. J., Law, M. R., Morris, J. K., Wald, D. S. 2001. Quantifying the effect of folic acid. *Lancet* 358(9298):2069-2073.

Wilkinson, J. M., Pollard, I. In utero exposure to caffeine causes delayed neural tube closure in rat embryos. *Teratog Carcinog Mutagen.* 1994;14(5):205-11.

Wilson, M. P., Hugge, C., Bielinska, M., Nicholas, P., Majerus, P. W., Wilson, D. B. 2009. Neural tube defects in mice with reduced levels of inositol 1,3,4-trisphosphate 5/6-kinase. *Proc. Natl. Acad. Sci. US* 106(24):9831-9835.

Wilson, M. P., Majerus, P. W. Isolation of inositol 1,3,4-triphosphate 5/6-kinase, cDNA cloning and expression of the recombinant enzyme. *J. Biol. Chem.* 1996 May 17;271(20):11904-10.

Yanfeng Wang, Lurong Lian, Jeffrey A. Golden, Edward E. Morrisey, Charles S. Abrams. PIP5KIγ is required for cardiovascular and neuronal development. *Proc. Natl. Acad. Sci. US.* 2007 July 10; 104(28): 11748–11753

In: Neural Tube Defects
Editor: Alberich Klein
ISBN: 978-1-62417-892-4

Chapter IV

Folic Acid, Homocysteine and Neural-Tube Defects

Yara Maria Rauh Müller**, *Karoline Kobus and Evelise Maria Nazari

[1]Universidade Federal de Santa Catarina, Centro de Ciências Biológicas, Programa de Pós-graduação em Biologia Celular e do Desenvolvimento, Campus Universitário, Florianópolis, SC, Brazil

Abstract

Knowledge of the cellular mechanisms responsible for normal neural-tube formation is essential to comprehending the genesis of neural-tube defects (NTD), as well as their prevention. Neurulation is the early embryonic process in which the neural plate undergoes shaping and folding to form the neural-tube, the precursor of the central nervous system (CNS). NTD result from a failure of the rise and fusion of the neural folds, which can affect the brain and/or spinal-cord. The mechanism of NTD genesis has not yet been established, but there is consensus in the literature that it involves disruptions in neural and also mesenchymal tissues. These defects have a multifactorial etiology, a combination of genetic predisposition and/or environmental factors. In humans, supplementation with folic acid (FA) during preconception and

* Corresponding author: Yara Maria Rauh Müller.

early pregnancy prevents these defects. Dietary FA deficiency results in increased levels of homocysteine (Hcy) in the plasma. Hcy is a sulfur-containing amino acid derived from the homocysteine-methionine cycle and from the transsulfuration pathway to cysteine. In the hyperhomocysteinemia condition, the reduced availability of FA compromises nucleic-acid synthesis, which is necessary to maintain the rapid and intense cell proliferation, as well as the synthesis and expression of the genes and proteins that occur during early embryogenesis. Experimental studies have focused on the teratogenic role of Hcy in different animal models, demonstrating its effects on the occurrence of congenital anomalies. The main aim of this chapter is to describe the effects of FA and Hcy imbalance on the cellular dynamics of CNS development, using embryos of *Gallus domesticus* as the experimental model. We showed that Hcy induces different disruptions in external morphology, mainly NTD. Regarding the cellular organization of the CNS, Hcy-treated embryos exhibited a reduction in the thickness of the neuron-glial layers. In summary, these results demonstrate that the imbalance between FA and Hcy levels can lead to disruptions in CNS and mesenchyme development, affecting proliferation and cell adhesion, and consequently changing the arrangement of the neuron-glial layers.

Introduction

Neurulation is an essential stage of vertebrate development, defined as a complex set of dynamic morphogenetic movements that result in formation of the neural-tube, the precursor of the central nervous system (CNS). Developmental biologists have investigated neurulation in different animal models, including fish, amphibian, chick, mouse, and also in humans. The stage of neurulation is a well-documented phenomenon, and knowledge about similarities and/or differences of cellular and molecular mechanisms involved in neural-tube formation, among different species, is necessary to fully decode the mechanisms that drive this embryonic process.

The neural-tube originates from two distinct morphogenetic processes, primary and secondary neurulation. In primary neurulation, the neural-tube initiates from the ectodermal neural plate and generates the brain and most of the spinal-cord; secondary neurulation is responsible for originating the caudal end of the spinal-cord, from a mesenchymal medullary cord derived from the tail bud [1, 2, 3].

Primary neurulation has been described in different species (chick, rabbit, frog, pig, mouse and humans), and in general, the basic steps of neural-tube formation are conserved among species [4, 5, 6, 7, 8, 9]. The shape and furrow of the neural plate are produced by intricate coordination of the neuroectoderm cells, which include changes in the morphology, position and number of cells in the future neural-tube architecture [10]. Molecular studies have demonstrated the dynamics during neurulation, which include, besides cell proliferation and differentiation, the cell polarity, adhesion and key signaling pathways [11, 12, 13]. Molecular interactions among the ectoderm, mesenchyme and notochord are necessary to express the neural fate in folding nd closure of the neural-tube [14, 15, 16].

In mice and in humans, morphological and histological evidence confirms that the cranial neural-tube closure is initiated at multiple sites [9, 17]. The closure is completed when the anterior and posterior openings of the neural-tube (anterior and posterior neuropores) close. Primary neurulation is completed when the posterior neuropore is closed, and then secondary neurulation begins [18, 19].

At the end of the neurulation process, the primary and secondary neurocoels become continuous and a single neural-tube extends from the head to the tail [20]. Interestingly, in chick embryos there is an overlap zone between primary and secondary neurulation at the level of the posterior neuropore. In the overlap region, primary neurulation occurs more dorsally and secondary neurulation more ventrally. Thus, in this region, there are two neural tubes that eventually coalesce into a single tube [21, 22].

Developmental biologists, epidemiologists and clinicians have focused their attention on the neurulation process, with the aim of understanding the origin of neural-tube defects (NTD), which include a variety of serious congenital anomalies.

In general, NTD result from a failure of the neural-tube to properly close during the earliest stages of development, and can affect the brain and/or spinal-cord [23]. Other tissues are also involved in the genesis of NTD. Disruptions in the differentiation of the cephalic and axial mesenchyme may cause anomalies in skeletal tissues associated with the CNS [24].

These complex congenital anomalies of the CNS are categorized clinically as “open,” in which the affected nervous tissues are exposed to the environment; or “closed”, in which the defect is covered by skin [25]. Open NTD is the most common and severe form of these anomalies, and includes anencephaly/exencephaly and myeloschisis/myelomeningoceles (open spina bifida), which result from the failure of fusion in the cranial and spinal region of the neural-tube, respectively [26]. The exact mechanism of NTD genesis has not yet been established, but there is a consensus in the literature that both categories, open and closed NTD, involve disruptions in neural and mesenchymal tissues.

Worldwide, it is estimated that 300,000 newborns are affected by anencephaly and/or spina bifida each year. Anencephalic newborns usually survive for only a few hours after birth. Children affected by spina bifida survive, but frequently have additional disorders that include a range of paralysis of the lower limbs, urinary and rectal incontinence, and less often, hydrocephaly [27].

In recent decades, the use of folic acid (FA), a water-soluble B9 vitamin, during preconception and early pregnancy to avoid disruptions in CNS development and to prevent the occurrence of human NTD has led to a renewed focus on neurulation [28, 29, 30, 31, 32]. In the following sections of this chapter, we discuss the effects of dietary FA deficiency and its impacts on the differentiation of the anterior and posterior regions of the neural-tube. We also present experimental data about the effects of FA deficiency on (i) external morphogenesis of embryos, using the chick as a model system for experimental embryology; and on (ii) the neural-cell dynamics during the differentiation of the brain and spinal-cord.

Folic Acid and Homocysteine Roles during Embryonic Development

The occurrence of NTD in humans is related to alterations in FA metabolism [33, 34, 35, 36]. In 1976, it was demonstrated that the diets and postpartum blood of women who had given birth to babies with NTD were deficient in FA [37]. Since 1991, the World Health Organization (WHO) and the US Food and Drug Administration (FDA) have recommended a daily intake of 400 μg of FA during conception up to the third month of gestation, i.e., the periconceptional period [38]. In fact, the use of FA is associated with a reduction in the risk of women having NTD-affected pregnancies [39, 40, 41, 42, 43]. Epidemiological studies have revealed that maternal ingestion of FA during the periconceptional period prevents the occurrence of NTD in about 50 to 70% of cases [44, 45, 46, 47].

FA deficiency results in increased levels of homocysteine (Hcy) in the plasma [48, 49, 50]. Hcy is a thiol (SH)-containing amino acid, which is not used in protein synthesis, but is involved at the connection of two major metabolic pathways. First, it can be condensed with the amino acid serine in an irreversible sulfuration reaction catalyzed by the enzyme cystathionine beta synthase in the presence of pyridoxine (vitamin B6) to form cystathionine. Hcy can also be reversibly methylated to re-form the essential amino acid methionine, taking methyl groups from the FA pathway in a reaction catalyzed by the enzyme methionine synthase in the presence of cobalamin (vitamin B12), thus allowing the further methylation processes that are vital to ongoing cellular activity of the living organism [51, 52].

In general, we highlighted two main processes in the metabolism of FA; first, the remethylation of Hcy to produce the essential amino acid methionine; and second, the synthesis of purine and thymidine required for DNA synthesis. Methionine, besides protein synthesis, is related to the formation of s-adenosylmethionine (AdoMet), which acts as a substrate for several transmethylation reactions, including DNA methylation. Interestingly, DNA methylation during the early stages of embryonic development promotes rapid changes in embryonic cells, which are involved with the transition from the undifferentiated to the differentiated state. The additional FA intake is very important for embryonic development, because it is able to prevent cellular disturbances related to DNA integrity, reducing the occurrence of NTD [36, 53].

High levels of Hcy tend to accumulate intracellularly and may be related to decreased DNA methylation, resulting in DNA breakage or triggering of abnormal apoptosis in the nervous system [54]. Deficient FA levels induce alterations in cytoskeletal proteins of embryos, and changes *in vitro* the differentiation of the neural crest and neuroepithelial cells, and decrease cell proliferation [55, 56, 57].

Experimental studies using different animal models have revealed that Hcy induces a significant decrease in embryonic survival rates [28, 58, 59, 60]. Additionally, studies have demonstrated that different chemical structures of Hcy decrease chick embryo survival; however, the results concerning embryonic survival after Hcy treatments are variable, and it is difficult to establish comparisons between these data, as illustrated in Table 1. This variability can partly be explained by differences in the time and duration of Hcy exposure, the chemical structure of Hcy used, the method employed for treatment, and the period between exposure and assessment of the morphological outcome.

Nevertheless, it is well accepted that Hcy significantly affects the chick-embryo survival rate, and as previously demonstrated, that this effect is dose-dependent [61].

Table 1. Survival comparative data obtained after performed Hcy treatment of the chick embryos

Chemical structure	Dose	Local of application	Period of application	Time of assessment	Survival rate	References
D,L-Hcy	5 – 7.5 μmol	droplet on the inner shell membrane	0 h, 24 h and 48 h	53 h	≥ 85%	Rosenquist et al. (1996)
	10 μmol				≥ 85%	
	15 μmol				≤ 25%	
	5 μmol	injection in the air sac of the egg	0 h, 24 h and 48 h	11 days	~ 50%	Miler et al. (2003)
	10 μmol				~ 30%	
	15 μmol				~ 40%	
L-Hcy thiolactone	3.5 μmol	droplet on the inner shell membrane	72 h	6 days	~ 50%	Limpach et al. (2000)
	4.5 μmol				~ 40%	
	5 μmol				< 20%	
	10 μmol	droplet on the top of embryo	29 h and 33 h	3 and $4^{1/2}$ days	100%	Epeldegui et al. (2002)

High plasma levels of Hcy have been related to the occurrence of congenital anomalies in different animal models. Among these anomalies, in addition to NTD, are heart defects, facial and eye defects, and limb malformations [28, 30, 62, 63, 64, 65].

Investigations need to focus on the cellular mechanisms whereby elevated levels of Hcy or decreased levels of methionine could cause NTD. Some hypotheses are: (i) Hcy *per se* is toxic during the neurulation process; (ii) decreased levels of methionine due to an impaired remethylation of Hcy to methionine will result in decreased levels of AdoMet, which is the general methyl donor in the human body and therefore could interfere with neurulation by an inadequate gene methylation and thus gene expression; and by inadequate amino acid methylation, which could interfere with, for example, microfilament synthesis; and (iii) increased Hcy levels will lead to an increased concentration of s-adenosylHcy (AdoHcy) [53].

The effects of FA deficiency on development have long been debated; however, few morphological aspects of nervous-system development have been explored.

Therefore, in the next section of this chapter, we present some data about experimental studies performed with chick embryos. The processes involved in neural-tube formation and early CNS development are basically similar in mammals and chick embryos [7, 66, 67, 68, 69]. *In ovo* development of avians appears to be an excellent model for developmental studies, because it occurs in the absence of maternal factors.

Moreover, the embryos can be easily accessed and manipulated at any time during development, without maternal surgical intervention. In our studies, we first mimicked the hyperhomocysteinemia condition, and we then performed different treatments with FA to investigate the roles of this vitamin in preventing or reversing the harmful effects of Hcy.

Effects of FA and Hcy on Development of the Chick Embryo Central Nervous System

Studies in birds and rodents have demonstrated a correlation between high Hcy levels and the occurrence of congenital anomalies [31, 64, 70]. We performed experiments[1] using 0.5 μg of FA and 10 and 20 μmol of D,L-Hcy, both diluted in saline solution. These doses of FA and Hcy were based on experiments concerning FA and Hcy *in ovo* exposures conducted by Rosenquist and col. [62] and on a previous dose-response study performed in our laboratory.

We treated the embryos at 24 and 48 hours of incubation, corresponding to the first and second embryonic days (E1 and E2). These ages were chosen because the neurulation process has not yet concluded. After treatment, the eggs were returned to the incubator (37.5 - 38.0°C and 65.0% relative humidity) until E4 and E6, when morphological and cellular analyses were performed.

We first examined the general quantitative parameters, including the body mass and cephalic measurements. The body mass did not differ among embryos-treated at both embryonic ages analyzed, as shown in Table 2. Two main measurements of the cephalic region of the chick embryos with normal external morphology were performed: (i) high cephalic length (HCL) and (ii) antero-posterior cephalic length (APCL). Morphometric analysis revealed that the treatment with AF or Hcy did not change the size of the encephalic vesicles of the chick embryos with normal external morphology, as seen in Figure 1.

---

[1] All experiments were carried out according to the guidelines of the institution's Ethics Committee (Protocol 254/CEUA/UFSC).

Table 2. Body mass of the control and FA and Hcy treated chick embryos

Embryonic age	control	FA	Hcy
E4	0.060 (± 0.005) (n = 32)	0.066 (± 0.006) (n = 35)	0.055 (± 0.004) (n = 39)
E6	0.42 (± 0.02) (n = 11)	0.44 (± 0.02) (n = 12)	0.39 (± 0.02) (n = 25)

Data are expressed as mean ± SEM.

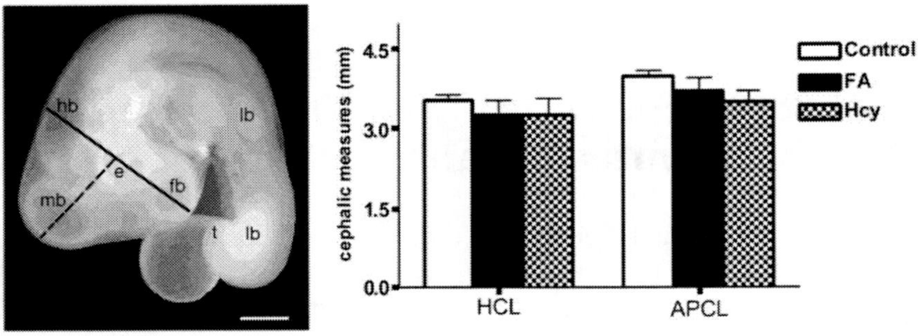

Figure 1. Chick embryo at E4, showing normal morphology. Dotted line represents the high cephalic length (HCL) and solid line represents the antero-posterior cephalic length (APCL), measured in the embryos treated with FA and Hcy in E4 and E6. e, eye; fb, forebrain; hb, hindbrain; lb, limb bud; mb, midbrain; t, tail. Scale bar: 4 mm. Modified from Kobus [65]. Accompanying graph shows the effect of the FA and Hcy treatments on cephalic measurements. Data are expressed as mean ± SEM.

The treatments with 10- and 20-μmol doses of Hcy at E1 and E2 significantly affected the external morphology of the chick embryos at E4 and E6, according to the data shown in Table 3 and Figure 2. In contrast, the control embryos did not exhibit disruptions of the external morphology at the same embryonic ages. Embryos at E4 treated with FA showed no DTN, only lumbar torsion. Embryos treated with Hcy, at both ages E4 and E6, exhibited DTN and also other congenital anomalies, including lumbar torsion, limb-bud malformations, and facial and eye defects.

The association between treatment with Hcy *in ovo* and NTD occurrence has been previously established by other authors. Rosenquist and col. [71] recorded an incidence of spina bifida of about 35% after Hcy treatment *in ovo* with a dose of 15 μmol of L-Hcy thiolactone.

Table 3. Morphological alterations observed in control and treated chick embryos

Morphological alterations		Embryonic ages					
		E4			E6		
		Frequency (%)					
		Control (n=32)	FA (n=35)	Hcy (n=39)	Control (n=11)	FA (n=12)	Hcy (n=25)
NTD	Exencephaly	0.0	0.0	7.7	0.0	0.0	12.0
	Spina bifida	0.0	0.0	25.6*	0.0	0.0	25.6*
Other anomalies	Lumbar torsion	0.0	5.7	28.2*	0.0	0.0	0.0
	Posterior limb buds defects	0.0	0.0	2.6	0.0	0.0	5.1
	Facial defects	0.0	0.0	2.6	0.0	0.0	7.7
	Asymmetric eyes	0.0	0.0	5.1	0.0	0.0	7.7
Total		0.0	5.7	74.4	0.0	0.0	68.4

Hcy = homocysteine treated embryos; FA = folic acid treated embryos. n = denotes the number of embryos analyzed. *denotes significant differences when compared to control group by Student-Neuman Keuls test ($p < 0.05$).

Epeldegui and col. [28] treated chick embryos with 20 μmol of L-Hcy thiolactone at stage 9+/10 HH (35 hours) and analyzed the external morphology at stage 21 HH (3½ days). These two research groups also found an incidence of exencephaly of about 16%. Other authors have reported the occurrence of congenital malformations, including NTD, craniofacial defects, reduced size of the eyes, heart defects, and abnormal development of the posterior limb bud in E6 chick embryos after *in ovo* treatment with Hcy [63]. Afman and col. [29] demonstrated that *in ovo* Hcy administration results in several malformations, including gross abnormalities, blisters, somite deformation, caudal degeneration, blood spots and disturbances in vein development of the chick; however, they did not observe increased NTD occurrence.

Another study [64] demonstrated that a dose of 20 μmol of L-Hcy thiolactone, administrated at 38 hours of incubation, is teratogenic for eye development, affecting the optic cup and lens of the chick embryos.

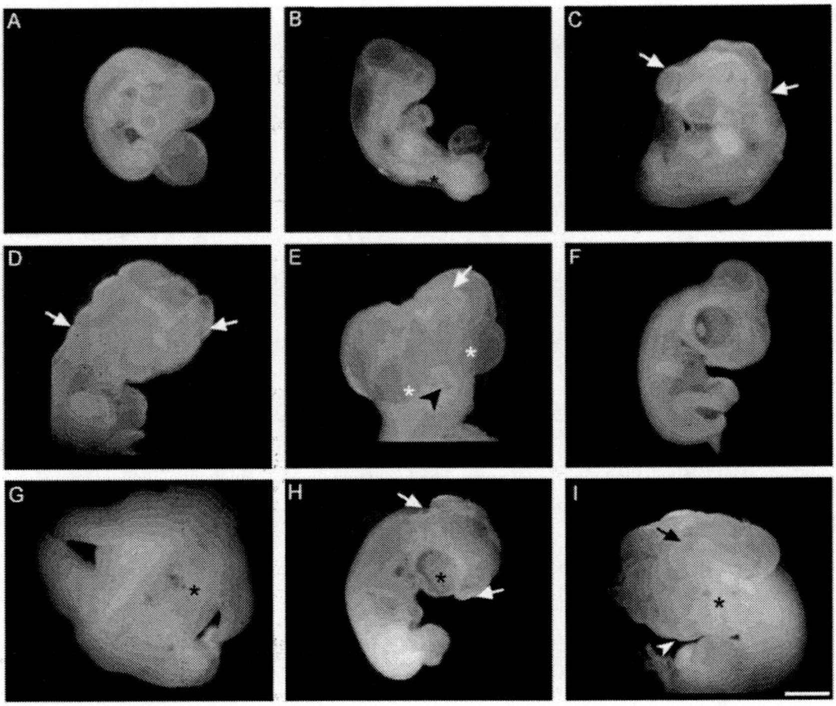

Figure 2. Morphological alterations induced by Hcy treatment in E4 and E6 chick embryos. (A) E4 control embryo, exhibiting normal morphology. (B) E4 Hcy-treated embryo, showing spina bifida in the lumbar region (black asterisk). (C) E4 Hcy-treated embryo, exhibiting excenphaly with open and everted neural folds extending from the forebrain to the midbrain boundary (white arrows). (D) E4 Hcy-treated embryo, exhibiting excenphaly with open and everted neural folds extending from the forebrain to the cervical/hindbrain boundary (white arrows). (E) E4 Hcy-treated embryo, showing exencephaly (white arrow), asymmetrical eye development (white asterisks) and facial defects (black arrowhead). (F) E6 control embryo, exhibiting normal morphology. (G) E6 Hcy-treated embryo, showing spina bifida in the thoraco-lumbar region (black asterisk). (H) E6 Hcy-treated embryo, exhibiting severe exencephaly extending from the forebrain to the cervical/hindbrain boundary (white arrows). (I) E6 Hcy-treated embryo, showings protrusion of an enlarged and morphologically abnormal brain (black arrow) and asymmetrical eye development (black asterisks) and facial defects (white arrowhead). Scale bar: 3 mm in A and C; 5.5 mm in B; 2.5 mm in D and E; 3.5 mm in F; 1 mm in G; 2 mm in H; 1.5 mm in I.

As mentioned previously, there were no changes in the dimensions (HCL and APCL) of the encephalic vesicles after Hcy and FA treatments. However, when we evaluated the cellular organization of the telencephalon, we identified a significant decrease in the thickness of the ventricular and mantle layers at E4 in chicks treated with Hcy, as shown in Figure 3.

Alterations in cell-proliferation dynamics is one probable reason for the decrease in telencephalon thickness after Hcy, mainly in the ventricular layer of the telencephalon. We demonstrated [65] that a dose of 10 μmol of Hcy is sufficient to sharply decrease the number of proliferating cells in the ventricular layer of the spinal-cord of embryos at E4. The Hcy-induced reduction of the mantle layer strongly indicates changes in cell migration and differentiation, as we observed also in the arrangement of the spinal-cord layers. The effect of the imbalance between FA and Hcy levels on cellular features of nervous-system development is incompletely described. Alterations in cytoskeletal proteins of the rat embryos after methionine deprivation in culture were demonstrated by Moephuli and col. [55]; changes in the *in vitro* differentiation of the neural crest and neuroepithelial cells were revealed by Boot and col. [56]; and a decrease in the proliferation of the progenitor cells associated with an increase in apoptosis of progenitor cells in the fetal mouse brain was reported by Craciunescu and col. [57].

In the spinal-cord of embryos at E4, only the thickness of the mantle layer was reduced due to Hcy treatment, as shown in Figure 4. In contrast, at E6 the thicknesses of all cell layers were decreased by Hcy treatment (Figure 5). At both embryonic ages, Hcy induced a decrease in the thickness of the mantle layer.

Therefore, considering the data obtained after measurement of the telencephalic and spinal-cord layers, we suggest that somehow the Hcy *in ovo* treatment changes the neuronal and glial cell migration and differentiation, modifying the initial organization of the neural layers in the chick embryo CNS. This hypothesis is supported by our previous data [65], which demonstrated that Hcy treatment is able to reduce the expression of the neural cell adhesion molecule (NCAM) in the spinal-cord. The low NCAM expression in the spinal-cord of the embryos, which could cause a decrease in the adhesiveness between neurons, also contributes to disruptions of the cellular organization of the neural cell layers. Additionally, changes in the arrangement of cell layers were always observed in embryos with the most severe categories of spina bifida, after treatment with Hcy.

Embryos treated with Hcy developed spina bifida, and the microscopic analyses revealed changes in the spinal-cord morphology of embryos at E4 and E6, as seen in Figures 4 and 5. The spinal-cord of embryos at E4 also showed abnormalities in the size of the lumen (neurocoel), as illustrated in Figures 4 G and H.

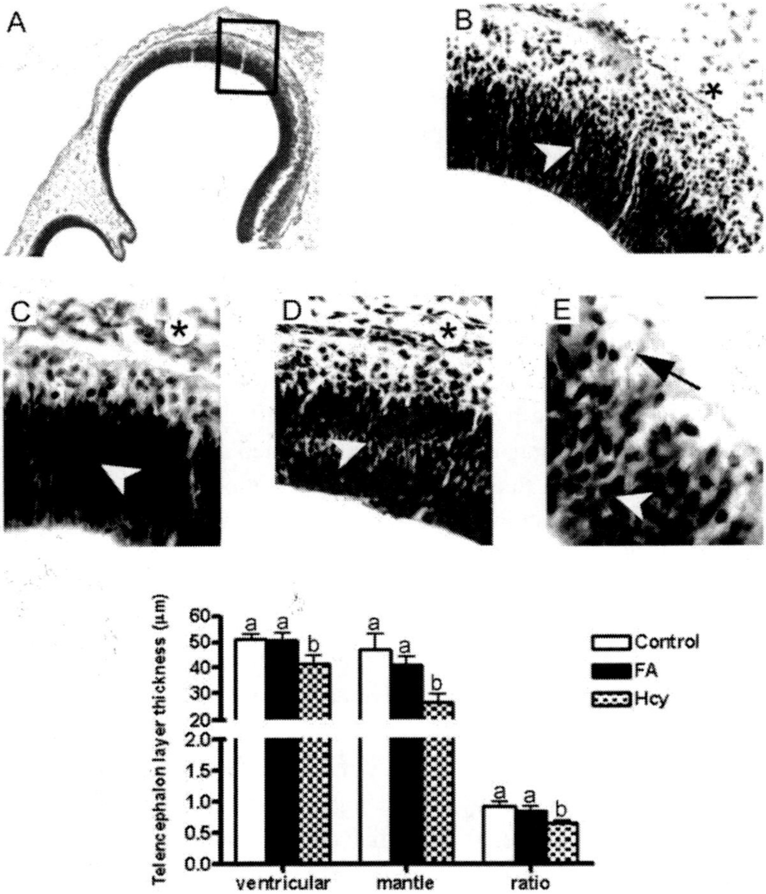

Figure 3. Frontal sections of telencephalon of control and treated embryos. (A) Telencephalon of control embryo showing, in square, the region where the telencephalon layers were measured. (B) Detail in high magnification of the square in A. (C) Embryo treated with FA. (D) Embryo treated with Hcy. (E) Higher magnification of the telencephalon layers of embryos treated with Hcy. White arrowheads show ventricular layers, asterisks show cephalic mesenchymal tissue, and black arrow indicates a decrease in the thickness of the mantle layer of the telencephalon of the embryo treated with Hcy. Scale bar: 190 µm in A; 40 µm in B; 30 µm in C; 25 µm in D; and 20 µm in E. Accompanying graph shows the effects of FA and Hcy on the ventricular and mantle layers in the telencephalon. Ratio corresponds to relationship between mantle and ventricular layers. Data are expressed as mean ± SEM. N = 18 measurements per treatment. Different letters represent a significant difference ($p < 0.05$) by one-way analysis of variance followed by Duncan's test.

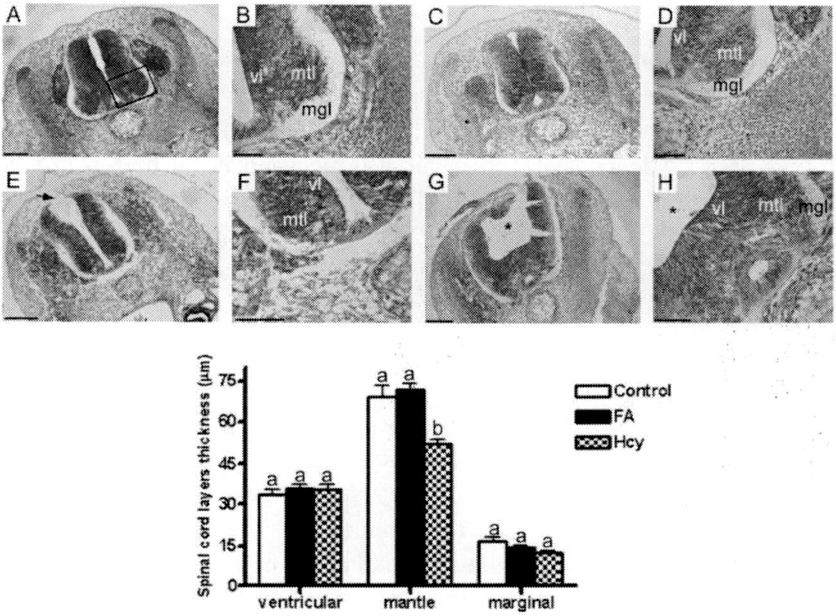

Figure 4. Transverse sections of the spinal-cord of the control and treated embryos at E4. Spinal-cord of the control (A) and FA-treated (C) embryos, exhibiting normal morphology. The square in A defines the region where the thickness of the spinal-cord layers was measured. High magnifications of the spinal-cord of the control (B) and FA-treated (D) embryos, indicating the position of the ventricular, mantle and marginal layers. (E) Spinal-cord of embryos treated with Hcy, exhibiting failure of neural-tube closure and non-neural ectoderm fusion (arrow). (F) Higher magnification of the spinal- cord shown in E, indicating the position of the ventricular and mantle layers. (G) Embryos treated with Hcy, exhibiting abnormal morphology of the spinal-cord with disruptions in the neurocoel morphology (asterisk). Higher magnification in H showing alterations in the arrangement of the cell layers in the abnormal spinal-cord. vl, ventricular layer; mgl, marginal layer; mtl, mantle layer. Scale bar: 100 µm in A, C, E and G; 50 µm in B, F, D and H. Accompanying graph shows the effect of FA and Hcy on the thickness of the ventricular, mantle and marginal layers of the spinal-cord of embryos at E4. Data are expressed as mean ± SEM. N = 34 in control; N = 46 in FA; N = 28 in Hcy. Different letters represent a significant difference ($p < 0.05$) by one-way analysis of variance followed by Tukey's test.

Additionally, at this same age, the embryos exhibited failure in neural-tube closure, accompanied by failure in non-neural ectoderm fusion. However, these embryos did not show disruptions in the arrangement of the spinal-cord layers (Figure 4 E, F). At E6, Hcy treatment induced spina bifida, with the presence of multiple neurocoels accompanied by a disorganized arrangement of the spinal-cord layers (Figure 5 G).

The disruptions in spinal-cord morphology of the embryos observed after Hcy treatment suggest the hypothesis that Hcy can alter the primary and secondary neurulation processes, although the underlying mechanisms remain unclear. It is evident that Hcy can modify mechanisms of both primary and secondary neurulation, which are distinct processes but overlap in some regions of the neural-tube in the chick [1, 22].

Conclusion

Since the 1970s, when Smithells and col. demonstrated that a FA dietary deficiency is related to the occurrence of NTD, a complex congenital anomalies of the nervous system, many studies have been performed with embryos in order to understand this finding. Besides NTD, other important congenital anomalies including facial, heart and limb defects have been related to FA deficiency.

In recent decades, it was established that FA deficiency increases the levels of amino acid Hcy in plasma. Hcy is extremely cytotoxic and induces many cell disruptions including DNA damage and abnormal apoptosis, impairs cytoskeletal proteins, and also affects the cell cycle. Hcy is also associated with decreased embryo survival and with the appearance of diverse congenital anomalies associated with different organs and systems.

Experimental studies that mimic the hyperhomocysteinemia condition revealed that Hcy induces significant macro- or micro-morphological disruptions in chick embryos. In spite of the benefits of FA in embryonic development, FA was not always able to neutralize the harmful effects caused by Hcy.

Considering that human embryos have less-efficient transsulfuration and remethylation abilities in comparison to the adult [72], also, chick embryos may not have had the competence to metabolize the FA injected, and consequently could not totally reverse the Hcy-induced harmful effects.

In this chapter, we presented experimental data that may contribute to better understanding the effects of FA and Hcy involved in brain and spinal-cord development. Actually, Hcy interfered significantly with the spinal-cord development of chick embryos, whereas FA *per se* did not produce morphological disruptions. Interestingly, in some experimental procedures, FA seems to reverse the harmful Hcy-induced effects [65].

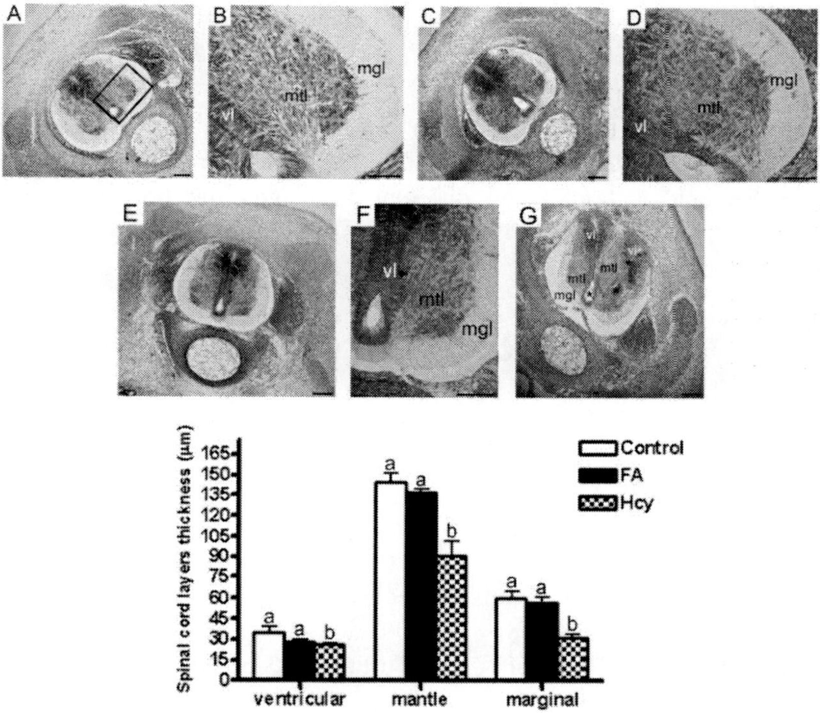

Figure 5. Transverse sections of the spinal-cord of the control and treated embryos at E6. Spinal-cord of the control (A) and FA-treated (C) embryos, exhibiting normal morphology. The square in A defines the region where the thickness of the spinal-cord layers was measured for all experimental groups. Higher magnifications of the spinal-cord of the control (B) and FA-treated (D) embryos, indicating the position of the ventricular, mantle and marginal layers. (E) Spinal-cord of the Hcy-treated embryo, exhibiting normal morphology of the spinal-cord. (F) Higher magnification of the spinal-cord shown in E, indicating the position of the ventricular, mantle and marginal layers. (G) Spinal-cord of the Hcy-treated embryo, exhibiting abnormal morphology of the spinal-cord, which contains multiple neurocoels (asterisks) and a disorganized arrangement of the spinal-cord layers. vl, ventricular layer; mgl, marginal layer; mtl, mantle layer. Scale bar: 100 µm in A, C, E and G; 50 µm in B and D; 75 µm in F. The accompanying graphs show the effects of the FA and Hcy treatments on the thickness of the ventricular, mantle and marginal layers in the spinal-cord of the control and treated embryos at E6. Data are expressed as mean ± SEM. N = 12 measurements for control group and N = 18 measurements for FA and Hcy groups. Different letters represent a significant difference ($p < 0.05$) by one-way analysis of variance followed by Tukey's test.

Complementary assays that answer the question as to whether cell matrix, adhesion molecules and neural crest-cell dynamics are involved in NTD must be performed to better comprehend the genesis of these defects and the role of FA in their prevention.

To conclude, NTD is an important worldwide health concern. Further studies are required to investigate whether the imbalance of FA and Hcy can change neuronal and glial cell differentiation and migration, and also to investigate the involvement of mesenchymal tissue in the genesis of NTD.

References

[1] Schoenwolf, G. C. Histological and ultrastructural observations of tail bud formation in the chick embryo. *Anat. Rec.* 1979, 193: 131–147.

[2] Schoenwolf, G. C., Smith, J. L. Mechanisms of neurulation: traditional viewpoint and recent advances. *Development* 1990, 109: 243–70.

[3] Lawson, A., Anderson, H., Schoenwolf, G. C. Cellular mechanisms of neural fold formation and morphogenesis in the chick embryo. *Anat. Rec.* 2001, 262: 153–168.

[4] Smith, J. L., Schoenwolf, G. C. Further evidence of extrinsic forces in bending of the neural plate. *J. Comp. Neurol.* 1991, 307: 225–236.

[5] Peeters, M. C., Viebahn, C., Hekking, J. W., Van Straaten, H. W. M. Neurulation in the rabbit embryo. *Anat. Embryol. (Berl.)* 1998, 197: 167–175.

[6] Davidson, L. A., Keller, R. E. Neural tube closure in *Xenopus laevis* involves medial migration, directed protrusive activity, cell intercalation and convergent extension. *Development* 1999, 126: 4547–4556.

[7] Van Straaten, H. W. M., Peeters, M. C. E., Hekking, J. W. M., Van der Lende, T. Neurulation in the pig embryo. *Anat. Embryo* 2000, 202: 75–84.

[8] Copp, A. J. Neurulation in the cranial region-normal and abnormal. *J. Anat.* 2005, 207: 623–635.

[9] Pyrgaki, C., Trainor, P., Hadjantonakis, A.-K., Niswander, L. Dynamic imaging of mammalian neural tube closure. *Dev. Biol.* 2010, 344: 941–947.

[10] Smith, J. L., Schoenwolf, G. C. Neurulation: coming to closure. *Trends Neurosci.* 1997, 20: 510–517.

[11] Ueno, N., Greene, N. D. E. Planar cell polarity genes and neural tube closure. *Birth Defects Res. C* 2003, 69: 318–324.

[12] Harris, M. J., Juriloff, D. M. Mouse mutants with neural tube closure defects and their role in understanding human neural tube defects. *Birth Defects Res. A* 2007, 79: 187–210.

[13] Copp, A. J., Greene, N. D. E. Neural tube defects – disorders of neurulation and related embryonic process. 2012. *WIREs Dev. Biol.* doi: 10.1002/wdev.71.

[14] Van Straaten, H. W. M., Hekking, J. W. M., Consten, C., Copp, A. J. Intrinsic and extrinsic factors in the mechanism of neurulation: effect of curvature of the body axis on closure of the posterior neuropore. *Development* 1993, 117:1163–1172.

[15] Wilson, P. A., Hemmati-Brivanlou, A. Vertebrate neural induction: inducers, inhibitors, and a new synthesis. *Neuron* 1997, 18: 699–710.

[16] Copp, A. J. Neurulation in the cranial region-normal and abnormal. *J. Anat.* 2005, 207: 623–635.

[17] Nakatsu, T., Uwabe, C., Shiota, K. Neural tube closure in humans initiates at multiple sites: evidence from human embryos and implications for the pathogenesis of neural tube defects. *Anat. Embryol* . 2000, 201:455–466.

[18] Nievelstein, R. A., Hartwig, N. G., Vermeij-Keers, C., Valk J. Embryonic development of the mammalian caudal neural tube. *Teratology* 1993, 48: 21–31.

[19] Lowery, L. A., Sive, H. Strategies of vertebrate neurulation and a re-evaluation of teleost neural tube formation. *Mechanisms of Development* 2004, 121: 1189–1197.

[20] Handrigan, G. R. Concordia discors: duality in the origin of the vertebrate tail. *J. Anat.* 2003, 202: 255–267.

[21] Criley, B. B. Analysis of the embryonic sources and mechanisms of development of posterior levels of chick neural tubes. *J. Morphol.* 1969, 128: 465–501.

[22] Schoenwolf, G. C., Delongo, J. Ultrastructure of secondary neurulation in the chick embryo. *Am. J. Anat.* 1980, 158: 43–63.

[23] O'Rahilly, R., Müller, F. Human embryology and teratology. 4 ed. New York: Wiley-Liss, Inc; 2001.

[24] Sadler, T. W. Embryology of neural tube development. *Am. J. Med. Genet. C* 2005, 135: 2–8.

[25] Suzuki, K. Neuropathology of developmental abnormalities. *Brain Dev.* 2007, 29: 129–141.

[26] Inagaki, T., Smith, J. L., Walker, M. L., Schoenwolf, G. C. Neural tube defects. *Methods Mol. Biol.* 2000, 136: 161–166.

[27] Christianson, A., Howson, C. P., Modell, B. March of Dimes Global Report on Birth Defects: the hidden toll of dying and disabled children. New York: March of Dimes Birth Defects Foundation, White Plains; 2006.

[28] Epeldegui, M., Peña-Melian, A., Varela-Moreiras, G., Pérez-Miguelsanz, J. Homocysteine modifies development of neurulation and dorsal root ganglia in chick embryos. *Teratology* 2002, 65: 171–179.

[29] Afman, L. A., Blom, H. J., Van der Put, N. M. J., Van Straaten, H. W. M. Homocysteine interference in neurulation: a chick embryo model. *Birth Defects Res. A* 2003, 67: 421–428.

[30] Brouns, M. R., Afman, L. A., Van Hauten, B., Hekking, J. W. M., Köhler, E. S., Van Straaten, H. W. M. Morphogenetic movements during cranial neural tube closure in the chick embryo and the effect of homocysteine. *Anat. Embryol.* 2005, 210: 81–90.

[31] Dunlevy, L. P. E., Burren, K. A., Chitty, L. S., Copp, A. J., Greene, N. D. E. Excess methionine suppresses the methylation cycle and inhibits neural tube closure in mouse embryos. *FEBS Letters* 2006, 580: 2803–2807.

[32] Abdul-Aziz, N., Turmaine, M., Greene, N. D. E., Copp, A. J. EphrinA-EphA receptor interactions in mouse spinal neurulation: implications for neural fold fusion. *Int. J. Dev. Biol.* 2009, 53: 559–568.

[33] Steegers-Theunissen, R. P., Boers, G. H., Trijbels, F. J., Eskes, T. K. Neural-tube defects and derangement of homocysteine metabolism. *New Engl. J. Med.* 1991, 324 (3): 199–200.

[34] Kirke, P. N., Mills, J. L., Scott, J. M. Homocysteine metabolism in pregnancies complicated by neural tube defects. *Nutrition* 1997, 13 (11–12): 994–995.

[35] Zhang, H-Y., Luo, G-A., Liang, Q-L., Wang, Y., Yang, H-H., Wang, Y-M., Zheng, X-Y., Song, X-M., Chen, G, Zhang, T., Wu, J-X. Neural tube defects and disturbed maternal folate- and homocysteine-mediated one-carbon metabolism. *Exp. Neurol.* 2008, 212: 515–521.

[36] Blom, H. J. Folic Acid, Methylation and neural tube closure in humans. *Birth Defects Res. A* 2009, 85: 295–302.

[37] Smithells, R. W., Sheppard, S., Schorah, C. J. Vitamin deficiencies and neural tube defects. *Arch. Dis. Child* 1976, 51: 944–950.

[38] MRC Vitamin Study Research Group. Prevention of neural tube defects: Results of the Medical Research Council Vitamin Study. *Lancet* 1991, 338: 131–137.

[39] Kalter, H. Teratology in the 20th century: environmental causes of congenital malformations in humans and how they were established. *Neurotoxicol. Teratol.* 2003, 25: 131–282.

[40] De Wals, P., Tairou, F., Van Allen, M. I., Uh, S. H., Lowry, R. B., Sibbald, B., Evans, J. A., Van den Hof, M. C., Zimmer, P., Crowley, M., Fernandez, B., Lee, N. S., Niyonsenga, T. Reduction in neural-tube defects after folic acid fortification in Canada. *New Engl. J. Med.* 2007, 357 : 135–142.

[41] Sayed, A. R., Bourne, D., Pattinson, R., Nixon, J., Henderson, B. Decline in the prevalence of neural tube defects following folic acid fortification and its cost-benefit in South Africa. *Birth Defects Res. A* 2008, 82: 211–216.

[42] Crider, K. S., Bailey, L. B., Berry, R. J. Folic acid food fortification – its history, effect, concerns, and future directions. *Nutrients* 2011, 3: 370–384.

[43] Safi, J., Joyeux, L., Chalouhi, G. E. Periconceptional folate deficiency and implications in neural tube defects. *J. Pregnancy* 2012. doi:10.1155/2012/295083.

[44] Daly, S., Millis, J. L., Molloy, A. M., Conley, M., Lee, Y. J., Kirke, P. N., Weir, D. G., Scott, J. M. Minimum effective dose of folic acid for food fortification to prevent neural-tube defects. *Lancet* 1997, 350: 1666–1669.

[45] Molloy, A. M. The role of folic acid in the prevention of neural tube defects. *Food Sci. and Tech.* 2005, 16: 241–245.

[46] Collins, J. S., Atkinson, K. K., Dean, J. H., Best, R. G., Stevenson, R. E. Long term maintenance of neural tube defects prevention in a high prevalence state. *J. Pediatr.* 2011, 159 (1): 143–149.

[47] Hursthouse, N. A., Gray, A. R., Miller, J. C., Rose, M. C., Houghton, L. A. Folate status of reproductive age women and neural tube defect risk: the effect of long-term folic acid supplementation at doses of 140 μg and 400 μg per day. *Nutrients* 2011, 3: 49–62.

[48] De la Calle, M., Usandizaga, R., Sancha, M., Magdaleno, F., Herranz, A., Cabrillo, E. Homocysteine, folic acid and B-group vitamins in obstetrics and gynaecology. *Eur. J. Obstet. Gyn. R. B.* 2003, 107: 125–134.

[49] Brauer, P. R., Tierney, B. J. Consequences of elevated homocysteine during embryonic development and possible modes of action. *Curr. Pharm. Des.* 2004, 10: 2719–2732.

[50] Pintó, X., Vilaseca, M. A., Balcells, S., Artuch, R., Corbella, E., Meco, J. F., Vila, R., Pujol, R., Grinberg, D. A folate-rich diet is as effective as folic acid from supplements in decreasing plasma homocysteine concentrations. *Int. J. Med. Sci.* 2005, 2: 58–66.

[51] Hague, W. M. Homocysteine and pregnancy. *Best Pract. Res. Cl. Ob.* 2003, 17 (3): 459–469.

[52] Blom, H. J., Shaw, G. M., Den Heijer, M., Finnell, R. H. Neural tube defects and folate: case far from closed. *Nature Rev. Neurosci.* 2006, 7: 724–731.

[53] Van der Put, N. M. J., Van Straaten, H. W. M., Trijbels, F. J. M., Blom, H. J. Folate, Homocysteine and Neural Tube Defects: An Overview. *Exp. Biol. M.* 2001, 226: 243–270.

[54] Mattson, M. P., Shea, T. B. Folate and homocysteine metabolism in neural plasticity and neurodegenerative disorders. *Trends in Neurosc.* 2003, 26: 137–146.

[55] Moephuli, S. R., Klein, N. W., Baldwin, M. T., Krider, H. M. Effects of methionine on the cytoplasmic distribution of actin and tubulin during neural tube closure in rat embryos. *PNAS US* 1997, 94: 543–548.

[56] Boot, M. J., Theunissen-Steegers, R. P. M., Poelmann, R. E., Van Iperen, L., Lindemans, J., Gittenberger-de Groot, A. C. Folic acid and homocysteine affect neural crest and neuroepithelial cell outgrowth and differentiation in vitro. *Dev. Dynam.*2003, 227: 301–308.

[57] Craciunescu, C. N., Brown, E. C., Mar, M. H., Albright, C. D., Nadeau, M. R., Zeisel, S. H. Folic acid deficiency during late gestation decreases progenitor cell proliferation and increases apoptosis in fetal mouse brain. *J. Nutrition* 2004, 134: 162–166.

[58] Van Aerts, L. A. G. J. M., Klaasboer, H. H., Postma, N. S., Pertijs, J. C., Copius Peereboom, J. H., Eskes, T. K., Noordhoek, J. Stereospecific *in vitro* embryotoxicity of L-homocysteine in pre- and post-implantation rodent embryos. *Toxicol. in Vitro* 1993, 7 (6): 743–749.

[59] Van Aerts, L. A. G. J. M., Blom, H. J., Deabreu, R. A., Trijbels, F. J., Eskes, T. K., Copius Peereboom- Stegeman, J. H., Noordhoek, J. Prevention of neural tube defects by and toxicity of L-homocysteine in cultured postimplantation rat embryos. *Teratology* 1994, 50 (5): 348–360.

[60] Miller, R. R., Leanza, C. M., Phillips, E. E., Blacquire, K. D. Homocysteine-induced changes in brain membrane composition correlate with increased brain caspase-3 activities and reduced chick embryo viability. *Comp. Biochem. Phys. t B* 2003, 136 (3): 521–532.

[61] Van Mil, N. H., Oosterbaan, A. M., Steegers-Theunissen, R. P. M. Teratogenicity and underlying mechanisms of homocysteine in animal models: A review. *Reprod. Toxicol.* 2010, 30: 520–531.

[62] Rosenquist, T. H., Ratashack, S. A., Selhub, J. Homocysteine induces congenital defects of the heart and neural tube: effect of folic acid. *PNAS US* 1996, 93: 15227–15232.

[63] Limpach, A., Dalton, M., Miles, R., Gadson, P. Homocysteine inhibits retinoic acid synthesis: a mechanism for homocysteine-induced congenital defects. *Exp. Cell Res.* 2000, 260: 166–174.

[64] De las Casas, M. C., Epeldegui, M., Tudela, C., Varela-Moreiras, G., Perez-Miguelsanz, J. High exogenous homocysteine modifies eye development in early chick embryos. *Birth Defects Res.* 2003, 67: 35–40.

[65] Kobus, K., Nazari, E. M., Müller, Y. M. R. Folic acid and homocysteine effects on cell mechanisms of the chicken spinal cord development. *Histochem.Cell Biol.* 2009, 132: 525–532.

[66] Copp, A. J., Brook, F. A., Estibeiro, P. J., Shum, A. S. W., Cockroft, D. L. The embryonic development of mammalian neural tube defects. *Prog. Neurobiol.*1990, 35 (5): 363–403.

[67] Morriss-Kay, G., Wood, H., Chen, W-H. Normal neurulation in mammals. *In: Ciba Foundation Symposium 181*. Chichester: Wiley, 1994; 51–69.

[68] Shum, A. S. W., Copp, A. J. Regional differences in morphogenesis of neuroepithelium suggest multiple mechanisms of spinal neurulation in the mouse. *Anat. Embryol.* 1996, 194: 65–73.

[69] Peeters, M. C. E., Viebahn, C., Hekking, J. W. M., Van Straaten, H. W. M. Neurulation in the rabbit embryo. *Anat. Embryol.* 1998, 197: 167–175.

[70] Afman, L. A., Blom, H. J., Drittij, M. J., Brouns, M. R., Van Straaten, HWM. Inhibition of transmethylation disturbs neurulation in chick embryos. *Dev. Brain Res.* 2005, 158: 59–65.

[71] Rosenquist, T. H., Schneider, A. M., Monaghan, D. T. N-methyl_D-aspartate receptor agonists modulate homocysteine-induced developmental abnormalities. *FASEB Journal* 1999, 13: 1523–1531.

[72] Björklund, N. K., Gordon, R. A hypothesis linking low folate intake to neural tube defects due to failure of post-translations methylations of the cytoskeleton. *Int. J. Dev. Biol.* 2006, 50: 135–141.

In: Neural Tube Defects
Editor: Alberich Klein
ISBN: 978-1-62417-892-4

Chapter V

Developmental Delay in Interneuron, Motor Neuron, Sensory Tract, and Synapse Formation in Chick Spina Bifida

Seiji Matsuda, Hiroaki Nabeka and Tetsuya Shimokawa
Department of Anatomy and Embryology, School of Medicine,
Ehime University, Shitsukawa, Toon City, Ehime, Japan

Abstract

Spina bifida aperta (SBA) is a congenital malformation of the spinal cord with complications such as spinal ataxia and bowel and bladder dysfunction. It is believed that a primary neurulation disorder with failed closure of the neural plate contributes to SBA. We developed a chick model of surgery-induced SBA that shows spinal ataxia after hatching. Incision over the length of seven somites caused SBA-like malformations in 100% of hatched chicks. This model may facilitate studies of the pathological mechanisms that lead to leg dysfunction in SBA chicks as follows. (1) Functionally, the SBA chicks exhibited leg muscle paralysis or imbalance between agonist and antagonist leg muscles. Lesions in these SBA chicks were located in the spinal segments that give rise to motor neurons innervating the dysfunctional muscles. (2) Histological analyses revealed fewer interneurons at the site of the lesion in SBA chicks than in normal chicks, but no difference was observed in the number of motor neurons. Leg dysfunction in this SBA model may be attributable to the smaller number of interneurons in the spinal segments that contain motor neurons innervating the dysfunctional muscle. (3) Immunohistochemical staining with a motor neuron marker showed that a defect in the production and elimination of excess motor neurons during the early developmental stages in SBA chicks may be involved in the pathological mechanism of the motor complications of this disease, although the final number of motor neurons was not different. (4) Histological analyses of spinal cords showed that the area of the dorsal funiculus was smaller in hatched SBA chicks than in the normal controls. Additionally, compared with normal controls, the SBA group had fewer nerve fibers in the dorsal funiculus. The ascending sensory nerve pathway was revealed by tracing the degenerated nerve fibers using osmification. In control chicks, degenerated sensory nerve fibers were observed only in the ipsilateral dorsal funiculus. In contrast, degenerated sensory nerve fibers were observed in all funiculi of SBA chicks. Consequently, fewer sensory nerve fibers ascended to the thoracic dorsal funiculus in the SBA chicks compared with normal controls. (5) Morphometric analyses of synapses and extracellular space on large montage electron micrographs of the spinal cord revealed a delay in synaptogenesis and tissue maturation in SBA chicks. The present results show a decrease in the number of interneurons and synapses and abnormal development in the motor neuron and sensory tracts of SBA chicks. These findings suggest that the complex neural networks controlling the muscles developed until the late gestational period and indicate the necessity for early therapy to prevent SBA-induced neurological deficits.

1. Introduction

Spina bifida aperta (SBA) occurs during the embryonic period and is one of the most common human congenital defects of the central nervous system. Depending on the position of the lesion, SBA leads to postnatal physical disabilities, including paralysis of the legs, lack of bowel and bladder control, and abnormalities in the hips, knees, and feet. Although the etiology of SBA is unknown, the pathogenic mechanism is generally thought to be a primary disorder of neurulation with failure of the neural plate to close (Copp et al., 1990).

In early human fetuses and in the mouse SBA model, the neural plate differentiates normally, except for neurulation (Keller-Peck and Mullen, 1997; Osaka et al., 1978; Patten, 1953). In contrast, remnants of the neural tissue at the lesion site exhibit severe traumatic and degenerative alterations during the late gestational and postnatal periods in human SBA (Cameron, 1956; Emery and Lendon, 1973). These observations led to the hypothesis that the unprotected spinal cord in embryos with SBA may be subjected to continuous, progressive secondary damage due to constant contact with amniotic fluid (Millicovsky and Lazar, 1995).

Mammalian models of SBA have been developed using surgical manipulation of fetuses after formation of the ventral arch and its surrounding tissue (Housley et al., 2000; Inagaki et al., 1997; Meuli et al., 1995a, b, 1996). Several experimental studies with a sheep SBA model have revealed that a substantial proportion of SBA-related neurological deficits may be attributable to progressive secondary destruction of the neural tissue due to abnormal exposure to amniotic fluid (Meuli et al., 1995a, b, 1996). In addition, these studies suggest that *in utero* surgical closure of the exposed cord during gestation may prevent the neurological dysfunctions at birth. Consequently, *in utero* surgery has been initiated as an innovative therapy for human fetuses with SBA (Adzick et al., 1998). However, several reports have indicated that *in utero* surgery did not preserve motor function in human patients with SBA (Hirose et al., 2001; Sival et al., 2004; Tulipan et al., 1999). Discrepancies among the follow-up studies of patients with SBA may be related to variability in the level of the spinal cord at which spinal neuron disruptions occur, particularly because there is a ventral–dorsal gradient of proteins within the embryonic neural tube that is involved in neural differentiation (Barth et al., 1999; Ericson et al., 1997; Nguyen et al., 2000; Tanabe and Jessell, 1996).

Advantages of using the chick, as compared with mammals, for a SBA model are that surgical manipulation can be carried out relatively easily during the neurulation period and that SBA-induced motor dysfunction in chicks is similar to that in humans because both are bipedal. Some reports have detailed the creation of avian embryos with SBA by surgical incision of the neural tube roof plate (Campbell and Sohal, 1990; Clark and Scothorne, 1990; Mann and Persaud, 1979; Sim et al., 1995). Clark and Scothorne (1990) described the incidence of SBA in 5-day-old chick embryos in which the roof plates had been incised over a length equivalent to seven somites at stages 12–18 of the Hamburger and Hamilton (1951) developmental table; they suggested that this developmental stage is a crucial determinant of the incidence of SBA in chick embryos.

In the present study, the disorder of muscles, numbers of neurons, development of motor neurons and synapses, and the sensory tract were compared between SBA and normal chicks.

2. Material and Methods

This study was carried out in compliance with the Guide for Animal Experimentation of Ehime University. Fertilized White Leghorn chicken eggs at stages 17–21 of Hamburger and Hamilton (1951) were used. The surgical manipulation was conducted according to a procedure described previously (Kinutani et al., 1986; Mominoki et al., 2006; Wang et al., 2011; Tsujimura et al., 2011). Briefly, the neural tube roof plate was incised longitudinally with a custom-made microknife by starting at the cranial margin level of the 26th somite, which forms the sixth (Th6) and seventh (Th7) thoracic segments, and extending caudally. The microknife was inserted into the neural tube to approximately half the depth of the tube, and the roof plate was incised. The shell window was closed with transparent adhesive tape after the incision, and the egg was placed back in the incubator.

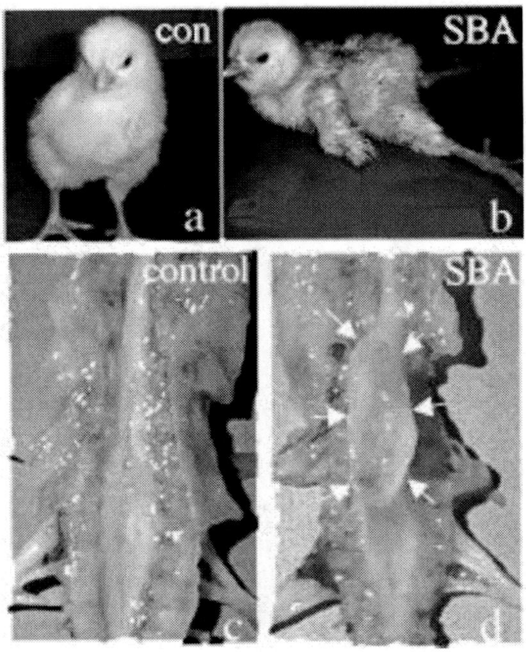

Figure 1. Locomotory behavior of normal control and spina bifida aperta (SBA) hatchlings (a, b). Normal control chicks stood soon after hatching (a). SBA chicks exhibited leg dysfunction and were unable to return a rotated thigh to the neutral position (b). Photographs of the dorsal view of the spinal cords of normal control (c) and SBA (d). The arrows indicate the position of the lesion (d).

3. Result

(1) Locomotory Behavior of SBA Hatchlings

Embryos at stages 18–19 were incubated until hatching. Hatchlings were kept in humidified incubators for 2 days without food and water. Locomotory behavior was recorded with a video-recorder for up to 2 days after hatching (Mominoki et al., 2006). Chicks with incomplete SBA and a five-somite lesion walked unsteadily. All SBA chicks with seven-somite lesions exhibited knee and ankle joint contractures (Figure 1b), despite slight differences in the location of the lesion, which extended from Th7 to L5 (Mominoki et al., 2006).

(2) Histological Observations of Spinal Neurons in the Lesioned Segment

The spinal cord between L2 and L4 was embedded in paraffin wax for sectioning, to quantify the number and size of neurons at the lesion site (Mominoki et al., 2006). Histological analyses revealed fewer small spinal neurons, most likely interneurons, at the level of the lesion in the SBA chicks compared with the normal control chicks. However, the number of large neurons did not differ between SBA and control chicks. The leg dysfunction that occurs in this model may be the result of a decrease in the number of spinal segment interneurons that modulate the motor neurons innervating the dysfunctional muscles (Mominoki et al., 2006).

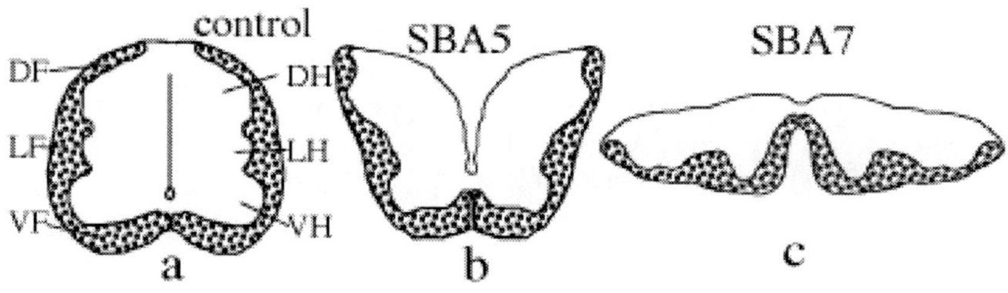

Figure 2. Schematic of transverse sections at the lumbar level of embryos from the control group (a), mild spina bifida aperta (SBA) chicks with a five-somite lesion (b), and SBA chicks with a seven-somite lesion (c).

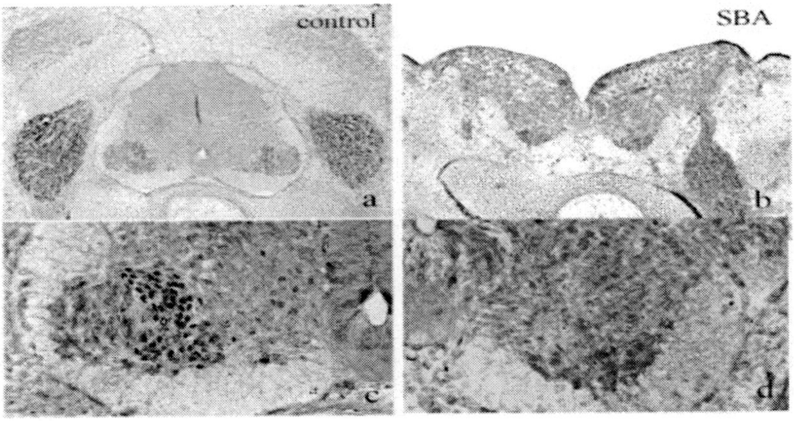

Figure 3. Distribution of islet-1-positive neurons in the neural tube at embryonic days (E) 6. Photomicrographs of transverse sections through the L3 neural tube in control (a, b) and spina bifida aperta (SBA) chicks (c, d) at E4.

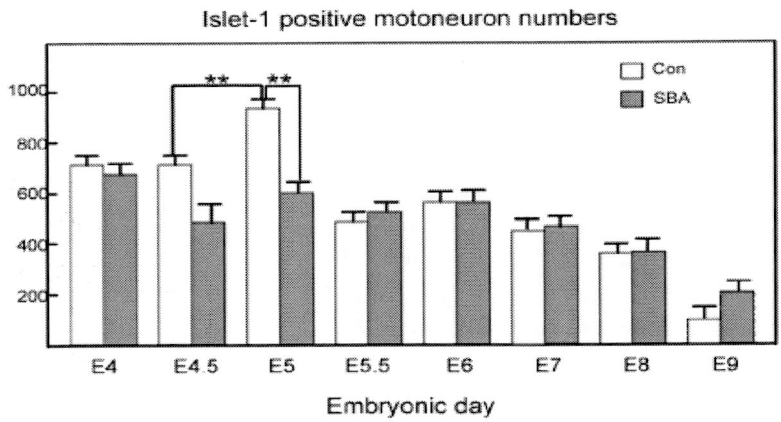

Figure 4. Chronological changes in the number of islet-1-positive neurons in the L3 ventral horn from embryonic day (E) 4 to E9 in control (white column) and spina bifida aperta (SBA) (stippled column) chicks.

(3) Developmental Motor Neuron Delay in SBA Chicks

Embryos were fixed with 4% paraformaldehyde and 0.2% glutaraldehyde, cryoprotected, and cryosectioned. Motor neurons in the early stages in chicks with and without SBA were observed by immunohistochemical staining with a monoclonal antibody against islet-1, a motor neuron marker (Figure 3; Wang et al., 2011).

Figure 4 shows the chronological changes in the islet-1 immunoreactive (islet-IR) cell numbers in the ventral horn in the control and SBA groups from embryonic day (E) 4 to E9. The islet-IR cell number in the control group increased abruptly at E5, decreased abruptly at E5.5, and then decreased gradually from E6 to E9. In contrast, islet-IR cell number in the SBA group decreased slowly, without any prominent increase at E5. The number of islet-IR cells in the SBA group was significantly lower than that in the control group from E4 to E5, particularly at E5. The number of islet-IR cells was similar between the groups after E6 (Figure 4; Wang et al., 2011).

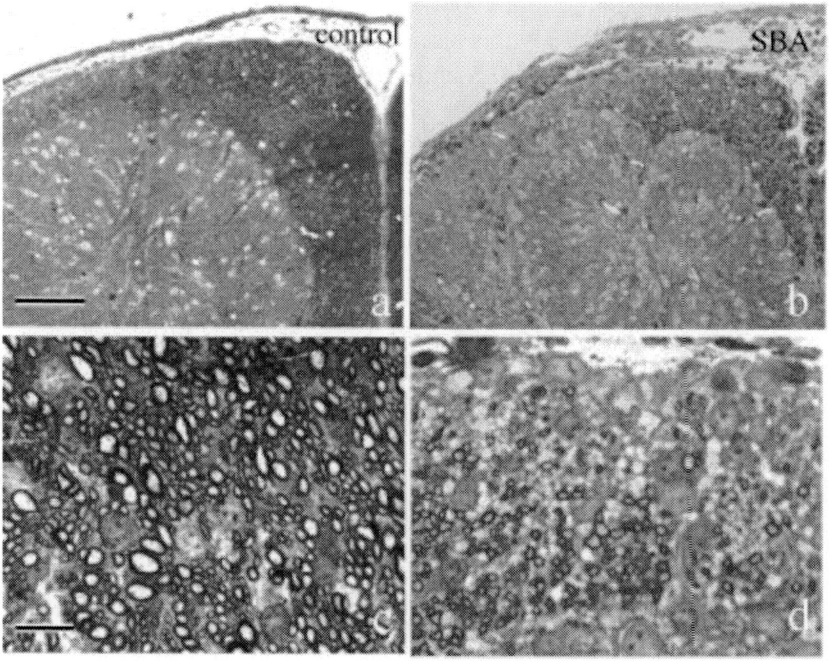

Figure 5. a, b: Transverse sections (toluidine-blue staining) through the seventh thoracic segment (Th7) of the dorsal funiculus in normal control (left column) and spina bifida aperta (SBA) (right column) chicks 4 days after hatching. The transverse area of the dorsal funiculus was significantly smaller in SBA chicks (b) than in normal control chicks (a). Scale bar indicates 50 μm. c, d: Photomicrographs of the dorsal funiculus are shown at high magnification. The density of nerve fibers was lower in the dorsal funiculus of the spinal cord in SBA chicks (d) than in normal control chicks (c). Scale bar indicates 10 μm.

Although the final numbers of islet-IR neurons in the two groups were not different, a defect in the production and elimination of excess motor neurons during the early developmental stages in the SBA group may be involved in the pathological mechanism of the motor complications in this disease.

(4) Tracing of the Sensory Tract

The area of the dorsal funiculus was smaller (Figure 5a, b), and nerve fibers ascending in the dorsal funiculus were clearly fewer in the SBA chicks compared with normal controls (Figure 5c–e). Previous studies have revealed no obvious decrease in the number of dorsal root ganglion sensory neurons in SBA chicks (Mominoki et al., 2006, Wang et al., 2011). Thus, we traced the sensory nerve fibers to clarify this discrepancy.

The dorsal root was cut at L5 under ether anesthesia 1 day after hatching. Three days after the neurotomy, the chicks were perfused transcardially with fixative and embedded in epoxy resin, and 0.5-μm-thick transverse sections were cut (Tsujimura et al., 2011). The osmiophilic degenerated axonal debris stained black in these histological sections (Figure 6). We traced the sensory fibers in the SBA (Figure 6e) and normal control chicks (Figure 6f).

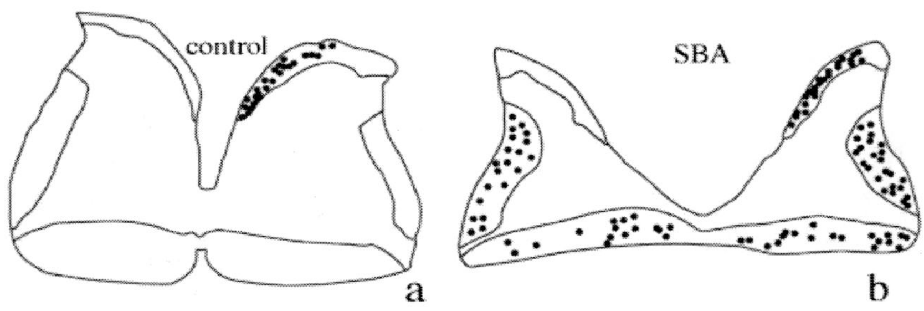

Figure 6. Degenerated nerve fibers (dots) observed at L5 in normal control (a) and spina bifida aperta (SBA) (b) chicks 3 days after neurotomy. In control chicks, degenerated nerve fibers were observed only in the ipsilateral dorsal funiculus, but not in the contralateral dorsal funiculus or lateral funiculus (b). In SBA chicks, degenerated nerve fibers were observed in the ipsilateral and contralateral dorsal and lateral funiculi (b).

Toluidine-blue-stained sections of the spinal cord fixed 3 days after neurotomy revealed that the dorsal funiculus of the control group contained many degenerated nerve fibers ipsilateral to the neurotomy, particularly near the dorsal median septum (Figure 6a, e). No degenerated nerve fibers were found in the lateral or ventral funiculus (Figure 6b, e). In contrast, degenerated nerve fibers were observed in the ipsilateral and contralateral dorsal (Figure 6b), lateral (Figure 6c), and ventral funiculi in SBA chicks.

Figure 7 is a schematic illustration of the course of the dorsal root ganglion neuron axons in the spinal cord of control (Figure 7a) and SBA chicks (Figure 7b). The spinal cord at the lesion site in SBA chicks exhibited a defect in the dorsal funiculus (Figure 2c; Mominoki et al., 2006), and the sensory nerve fibers did not ascend into the dorsal funiculus, but dispersed instead (Figure 7).

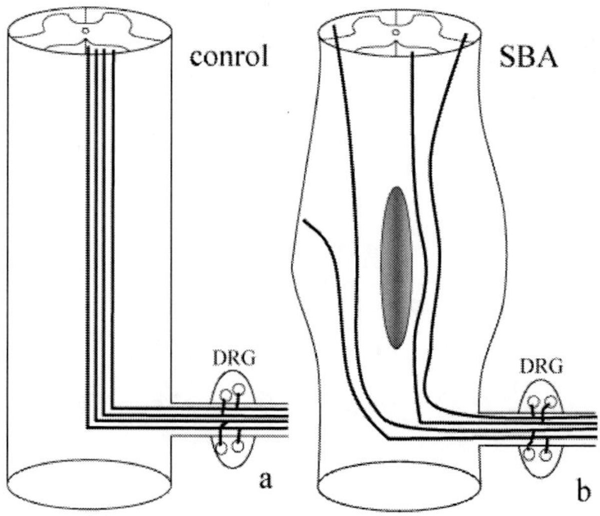

Figure 7. Schematic illustration of the course of sensory fibers from the dorsal root ganglion neurons (DRG) in the spinal cord of control (a) and spina bifida aperta (SBA) chicks (b).

(5) Morphometric Analysis of the Extracellular Space and Synapses by Large Montage Electron Microscopy

The chicks were anesthetized and perfused transcardially with a fixative. The spinal cord was embedded in epoxy resin, and ultrathin sections were mounted on single-slot grids coated with Formvar film using conventional electron microscopy methods. One photomontage of electron micrographs covering one-half of the spinal cord was prepared at a final magnification of ×3000.

Extracellular spaces >100 nm were traced and painted in black on a sheet of transparent paper (Sano et al., 1988). The total areas of the expansion as percentages of the whole cross-sectional area of the cervical spinal cord were 11.0, 7.7, 7.8, and 9.7% at stages 9, 11, 15, and 17, respectively. The area percentage decreased progressively thereafter, to 8.7, 3.0, 1.9, and 1.3% at stages 19, 22, 25, and 29, respectively. Extracellular space in the central portion of the lumbar cord at stage 31 was 1.9% in normal chicks and 14.9% in SBA chicks (Figure 8a–c).

Electron microscopic profiles were identified as synaptic boutons only when the profile had all of the characteristics of synaptic structures, including synaptic vesicles, synaptic density, and typical membrane apposition (Matsuda and Kanemitsu, 1985; Matsuda et al., 2004). The numbers of synapses in the cervical cord half were 2, 19, 57, 107, and 237 at stages 22, 25, 27, 29, and 31, respectively (Kanemitsu and Matsuda, 1985). The number of synapses in the SBA lumbar cord (28) was about half of the number in normal chicks (49) at stage 31 (Figure 9).

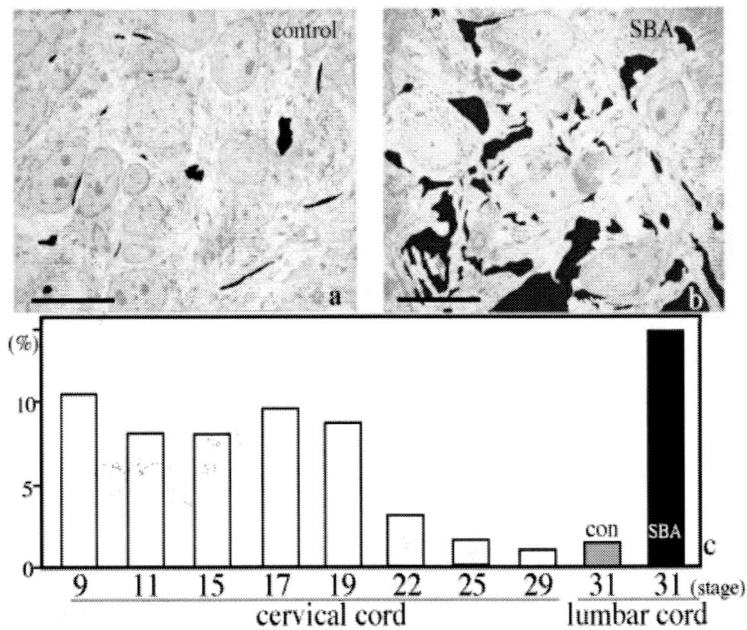

Figure 8. a, b: Cross-section of the stage 31 chick embryo lumbar spinal cord of control (a) and spina bifida aperta (SBA) chicks (b) showing the extracellular space painted in black. c: Changes in the percentage of extracellular space at each developmental stage.

only on the lateral side of the spinal cord (Mominoki et al., 2006). As a result, the dorsal funiculus, caudal or rostral to the site of the lesion, also developed abnormally. This may explain why the sensory nerve fibers in the SBA spinal cord do not ascend into the dorsal funiculus, but disperse in the lateral and ventral funiculi on both the ipsilateral and contralateral sides (Figures . 6, 7).

Sival et al. (2004) reported that the disappearance of leg movements is caused by lower motor dysfunction in the reflex arc, whereas neural descending conduction through the lesion site remains functional. Therefore, it is plausible that the motor conduction descending into the lateral funiculus is better preserved than the sensory conduction ascending into the dorsal funiculus. In the present study, we observed abnormal development and abnormal sensory nerve projections. We reported previously that fewer spinal interneurons occur at the lesion site in SBA chicks compared with control chicks and that no difference in the number of large motor neurons is observed (Mominoki et al., 2006). These findings suggest that leg motor dysfunction in our SBA model is not attributable to motor neuron damage or descending motor conduction, but rather to dysfunction of the neural circuit networks around the injured portion, such as interneurons and sensory input. Consistent with this hypothesis, abnormal shank dorsiflexion was exacerbated after the dorsal fibers were cut (Tsujimura et al., 2011). The deterioration was not observed in normal control chicks, even after dorsal root neurotomy. Although the mechanism underlying this phenomenon cannot be explained, the motor dysfunction can be attributed to the abnormal neural network in the SBA spinal cord.

Several experimental studies using a sheep SBA model have led to the hypothesis that a significant portion of the neurological deficits found in SBA may be caused by progressive secondary destruction of neural tissue due to abnormal exposure to amniotic fluid, and that *in utero* surgical closure of the exposed cord during gestation may prevent neurological dysfunctions at birth (Meuli et al., 1995a, b, 1996). As a result, *in utero* surgery to prevent neurological deficits has been initiated as an innovative therapy for human SBA (Adzick et al., 1998). However, some reports have indicated that *in utero* surgery does not necessarily preserve motor function in human patients with SBA (Hirose and Farmer, 2009; Hirose et al., 2001; Sival et al., 2004; Tulipan et al., 1999).

We showed a decrease in the number of interneurons (Mominoki et al., 2006), abnormal motor neuron development (Wang et al., 2011), an abnormal sensory tract (Tsujimura et al., 2011), and a decrease in the number synapses in our chick SBA model. These findings suggest that the complex neural circuit networks controlling muscles develop until the late gestational period and may explain why *in utero* surgical closure of the exposed spinal cord during this period is often insufficient to prevent SBA-induced neurological deficits. Our results suggest that early therapy continuing until this period is necessary to prevent SBA-induced neurological deficits.

Ackowledgments

The authors thank Dr. K. Mominoki, M. Wang, and R. Tsujimura for their valuable help. This study was supported in part by grants to SM from the Ministry of Education, Culture, Sports, Science, and Technology of Japan (19659380, 22591637) and from the Japan Spina Bifida and Hydrocephalus Research Foundation.

References

Adzick, N.S., Sutton, L., Crombleholme, T.M., Flake, A.W., 1998. Successful fetal surgery for spina bifida. *Lancet* 352, 1675–1676.

Barth, K.A., Kishimoto, Y., Rohr, K.B., Seydler, C., Schulte-Merker, S.,Wilson, S.W., 1999. Bmp activity establishes a gradient of positional information throughout the entire neural plate. *Development* 126, 4977–4987.

Cameron, A.H., 1956. The spinal cord lesion in spina bifida cystica. *Lancet* 271, 171– 174.

Campbell, L.R., Sohal, G.S., 1990. The pattern of neural tube defects created by secondary reopening of the neural tube. *J. Child. Neurol.* 5, 336– 340.

Clark, B.J., Scothorne, R.J., 1990. Variation in the response of chick embryos to incision of the roof plate of the neural tube at different developmental stages. *J. Anat.* 168, 167–184.

Copp, A.J., Brook, F.A., Estibeiro, J.P., Shum, A.S., Cockroft, D.L., 1990. The embryonic development of mammalian neural tube defects. *Prog. Neurobiol.* 35, 363–403.

Emery, J.L., Lendon, R.G., 1973. The local cord lesion in neurospinal dysraphism (meningomyelocele). *J. Pathol.* 110, 83–96.

Ericson, J., Rashbass, P., Schedl, A., Brenner-Morton, S., Kawakami, A., van Heyningen, V., Jessell, T.M., Briscoe, J., 1997. Pax6 controls progenitor cell identity and neuronal fate in response to graded Shh signalling. *Cell* 90, 169–180.

Guha, U., Gomes, W.A., Samanta, J., Gupta, M., Rice, F.L., et al., 2004. Target-derived BMP signalling limits sensory neuron number and the extent of peripheral innervation in vivo. *Development* 131, 1175–1186.

Hamburger, V., Hamilton, H.L., 1951. A series of normal stages in the development of the chick embryo. J. Morphol. 88, 49–92, Reprinted from: *Dev. Dyn.* 195, 231–272 (1992).

Hirose, S., Farmer, D.L., 2009. Fetal surgery for myelomeningocele. *Clin. Perinatol. 36*, 431–438.

Hirose, S., Farmer, D.L., Albanese, C.T., 2001. Fetal surgery for myelomeningocele. *Curr. Opin. Obstet. Gynecol.* 13, 215–222.

Housley, H.T., Graf, J.L., Lipshultz, G.S., Calvano, C.J., Harrison, M.R.,Farmer, D.L., Jennings, R.W., 2000. Creation of myelomeningocele in the fetal rabbit. *Fetal Diagn. Ther.* 15, 275– 279.

Inagaki, T., Schoenwolf, G.C., Walker, M.L., 1997. Experimental model:change in the posterior fossa with surgically induced spina bifida aperta inmouse. *Pediatr. Neurosurg.* 26, 185– 189.

Kanemitsu, A. Matsuda S. Growth of the initial dendrite of early differentiated interneurons in the zona intermedia of the embryonic chick cervical cord: A morphometric study with the Golgi method. *Neurosci Lett* 53:267-272, 1985

Keller-Peck, C.R., Mullen, R.J., 1997. Altered cell proliferation in the spinal cord of mouse neural tube mutants curly tail and Pax3 splotch-delayed. *Dev. Brain Res.* 102, 177–188.

Kinutani, M., Coltey, M., Le Douarin, N.M., 1986. Postnatal development of a demyelinating disease in avian spinal cord chimeras. *Cell* 45, 307–314.

Liem Jr., K.F., Tremml, G., Jessell, T.M., 1997. A role for the roof plate and its resident TGFbeta-related proteins in neuronal patterning in the dorsal spinal cord. *Cell* 91, 127–138.

Lo, L., Dormand, E.L., Anderson, D.J., 2005. Late-emigrating neural crest cells in the roof plate are restricted to a sensory fate by GDF7. *PNAS* 102, 7192–7197.

Mann, R.A., Persaud, T.V., 1979. Morphology of experimental spina bifida in the chick embryo. *Anat. Anz.* 145, 182–191.

Matsuda, S., Kanemitsu, A., Distribution of synapses in the upper cervical cord of chick embryo from stage 28 to 38. *Acta Anat Nippon.* 60: 361, 1985

Matsuda, S., Kobayashi, Y., Ishizuka, N., A quantitative analysis of the laminar distribution of sunaptic boutons in field CA3 of the rat hippocampus, *Neuroscience Res.*49: 241-252, 2004

Matsuda, S., Kobayashi, Y., Ishizuka, N., Dendro- and Synapto-Architecture of the Field CA3 of the Rat Hippocampus, *Neural Synapse Research Trends*, NOVA Publisher, 253-279, 2006

Meuli, M., Meuli-Simmen, C., Hutchins, G.M., Yingling, C.D., Hoffman, K.M., Harrison, M.R., Adzick, N.S., 1995a. In utero surgery rescues neurological function at birth in sheep with spina bifida. *Nat. Med.* 1, 342–347.

Meuli, M., Meuli-Simmen, C., Yingling, C.D., Hutchins, G.M., Hoffman, K.M., Harrison, M.R., Adzick, N.S., 1995b. Creation of myelomeningocele in utero: a model of functional damage from spinal cord exposure in fetal sheep. *J. Pediatr. Surg.* 30, 1028–1033.

Meuli, M., Meuli-Simmen, C., Yingling, C.D., Hutchins, G.M., Timmel, G.B., Harrison, M.R., Adzick, N.S., 1996. In utero repair of experimental myelomeningocele saves neurological function at birth. *J. Pediatr. Surg.* 31, 397–402.

Millicovsky, G., Lazar, M.L., 1995. Spina bifida: role of neural tissue damage during pregnancy in producing spinal paralysis. *Obstet. Gynecol.* 86, 300–301.

Mominoki, K., Kinutani, M., Kobayashi, N., Fujiwara, T., Matsuda, S., 2006. Leg dys- functions in a hatched chick model of spina bifida aperta. *Exp. Neurol.* 197, 133–142.

Nguyen, V.H., Trout, J., Connors, S.A., Andermann, P., Weinberg, E., Mullins, M.C., 2000. Dorsal and intermediate neuronal cell types of the spinal cord are established by a BMP signaling pathway. *Development* 127, 1209–1220.

Osaka, K., Tanimura, T., Hirayama, A., Matsumoto, S., 1978. Myelomeningocele before birth. *J. Neurosurg.* 49, 711–724.

Patten, B.M., 1953. Embryological stages in the establishing of myeloschisis with spina bifida. *Am. J. Anat.* 93, 365–395.

Sano, N., Matsuda S., Matsuda, H., Uehara, Y., A morphometric analysis of extracellular space in the developing spinal cord of the chick embryo. *Brain Res* 455:324-331, 1988

Schoenwolf, G.C., Desmond, M.E., Timing and positioning of reopening of the occluded spinal neirocele in the chick embryo. *J Comp. Neurol.,* 246:459-466, 1986

Sim, K.B., Cho, B.K., Chi, J.G., Wang, K.C., 1995. Morphological study of surgically induced open neural tube defect in old (14 and 21 days) chick embryos. *Neurosci. Lett.* 192, 61–64.

Sival, D.A., van Weerden, T.W., Vles, J.S., Timmer, A., den Dunnen, W.F., et al., 2004. Neonatal loss of motor function in human spina bifida aperta. *Pediatrics* 114, 427–434.

Stiefel, D., Copp, A.J., Meuli, M., 2007. Fetal spina bifida in a mouse model: loss of neural function in utero. *J. Neurosurg.* 106 (3 Suppl.), 213–221.

Tanabe, Y., Jessell, T.M., 1996. Diversity and pattern in the developing spinal cord. *Science* 274, 1115–1123.

Tsujimura, R., Mominoki K., Kinutani, M., Shimokawa, T., Doihara T, Nabeka, H., Wakisaka, H., Kobayashi, N., Matsuda, S., Sensory tract abnormality in the chick model of spina bifida, *Neurosci Res,* 71:85-91, 2011

Tulipan, N., Bruner, J.P., Hernanz-Schulman, M., Lowe, L.H., Walsh, W.F., Nickolaus, D., 1999. Effect of intrauterine myelomeningocele repair on central nervous system structure and function. *Pediatr. Neurosurg.* 31, 183–188.

Wang, M., Mominoki, K., Kinutani, M., Wang, Z., Kobayashi, N., et al., 2011. Chrono- logical changes in Islet-1-positive neurons in a chick model of spina bifida aperta. *J. Vet. Med. Sci.* 73, 447–452.

In: Neural Tube Defects
Editor: Alberich Klein

ISBN: 978-1-62417-892-4

Chapter VI

Do Neural Tube Defects Lead to Structural Alterations in the Urinary and Genital Organs in Human Fetuses?

Luciano Alves Favorito*
Urogenital Research Unit from State University
from Rio de Janeiro – Brazil

Abstract

Purpose: Anencephaly is the most severe neural tube defect in human fetuses. The objective of this paper is to analyze the structure of the ureter, bladder and penis in anencephalic human fetuses.

Methods: We studied 40 normal human fetuses (20 male and 20 female, aged 14 to 23 WPC) and 12 anencephalic fetuses (5 male and 7 female, aged 18 to 22 WPC). The ureter, bladder and penis were removed and processed by routine histological techniques. Stereological analysis of collagen, elastic system fibers and smooth muscle was performed in sections. Data were expressed as volumetric density (Vv-%). The images were captured with Olympus BX51 microscopy and Olympus DP70

* Correspondence Address: Luciano Alves Favorito MD, PhD Rua Professor Gabizo 104/201 - Tijuca - Rio de Janeiro -RJ – Brazil, Zip Code: 20271-062, Telephone number: 55(21) 25876499, Fax number: 55(21) 38728802/25876121, e-mail: lufavorito@yahoo.com.br

camera. The stereological analysis was done using the software Image Pro and Image J. For biochemical analysis, samples were fixed in acetone, and collagen concentrations were expressed as micrograms of hydroxyproline per mg of dry tissue. Means were statistically compared using the unpaired t-test ($p < 0.05$).

Results: We observed a significant increase ($p<0.0001$) in the Vv of collagen in the bladders of anencephalic fetuses (69.71%) when compared to normal fetuses (52.74%), and a significant decrease ($p<0.0001$) in the Vv of smooth muscle cells in the bladders of anencephalic fetuses (23.96%) when compared to normal fetuses (38.35%). The biochemical analyses showed a higher concentration of total collagen in the bladders of anencephalic fetuses (37354μg/mg) when compared to normal fetuses (48117μg/mg, $p<0.02$). Intracavernosal septum was present in all samples. We did not observed differences in the Ad of Cot and SMC in the penises of anencephalic fetuses when compared to normal ones. Performing simple Linear Regression, it can be suggested that during the human development, there is a gradual increase in Cot ($R^2= +0.45$) and a decrease in SMC ($R^2=-0.62$) in both groups studied. Elastin was observed only in fetuses from 20th WPC. There was no difference between gender, anencephalic and control groups, when calculating age, weight, ureteral wall thickness and lumen area. Using simple linear regression we could observe that ureteral lumen area ($R^2 = 0.52$) and wall thickness ($R^2=0.73$) increases gradually with ages advance, independent of the presence of anencephaly. This relation was not observed when using weight as a parameter ($R^2= 0.31$).

Conclusions: The structural alterations of the bladder found in this study may suggest the existence of functional alterations in the bladder of anencephalic human fetuses. There was no difference in the structure of the genitalia of anencephalic fetuses compared to normal ones. Elastin is documented from 20th WPC wich suggests the maintenance of erectile function. Histochemistry and immunolabeling suggests that penile shaft development is maintained and unaltered in NTD fetuses. Although anencephaly is a disease that is incompatible with babies' long spam survivors, this present study shows that ureteral development is unchanged. In our opinion, its anatomy did not suffer any alteration in consequence of this disease and, in the future, this tissue could be used as a substitute for native tissue.

Introduction

Neural tube defects (NTD) are one of the commonest congenital malformations of the central nervous system, with an average prevalence at birth of 1 per 1000 [1]. Anencephaly is the most severe of fetal NTD, resulting from failure of the neural tube closing at the base of the skull between the third and fourth weeks post-conception (WPC), leaving the skull bones that usually surround the head unformed. Thereby, the brain lacks part of the entire cerebrum, and the remaining brain tissue is often exposed to injury from the amniotic fluid [2].

Anencephaly invariably is lethal, but some anencephalic infants are born alive with a rudimental brain. The absence of a functional brain makes the anencephalic incapable of consciousness or feeling pain, but the brainstem reflexes can cause actions like breathing and, occasionally, they respond to sound and touch. Anencephalic newborns are not viable or tractable and the survival is measured in hours or days [3].

The pathogenesis of anencephaly is still controversial. Either a failure of closure of the neural tube or reopening after closure has been hypothesized. Several studies have suggested that anencephaly arises from exencephaly, in which the cerebral tissue not covered by the meninges, the cranium and skin is progressively destroyed within the uterus by the amniotic fluid [4].

Despite ethical conflicts, the literature shows some reports about the use of anencephalic fetuses' organs for transplantation [5-8]. The organ structure of anencephalic fetuses and children is almost unknown.

The potential use of tissue engineering using penile shaft from anencephalic fetuses would be useful in developing countries where penile cancer incidence and treatment increases [9] and in cases of pelvic trauma [10]. In addition to providing important information for experiments of the pathophysiology of penile disorders, data on penile extra-cellular matrix might also be beneficial for investigations of techniques of surgical reconstruction of the penis, especially those using different materials.

In this chapter we studied the structure of urinary (bladder and ureter) and genital (urethra and penis) organs in anencephalic fetuses and compare to normal ones.

Methods

The present work received institutional review committee and parent approval. This work was carried out in accordance to the ethical standards of the responsible institutional committee on human experimentation.

We studied 40 normal human fetuses (20 male and 20 female) and 12 anencephalic human fetuses (5 male and 7 female) without anomalies of the genitourinary tract (Figure 1).

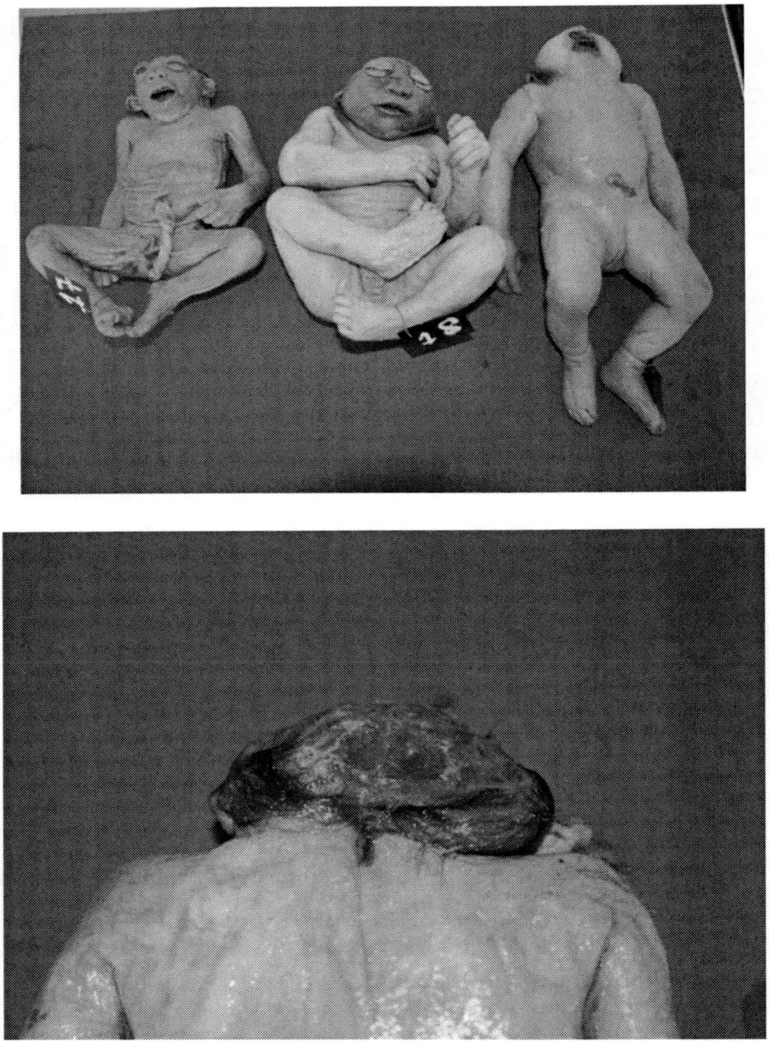

Figure 1. The figure shows the well-preserved anencephaly fetuses. A) Three fetuses in anterior view with anencephaly. B) Posterior view of a fetus with 23 weeks post-conception with anencephaly.

The gestational age of the fetuses was determined in weeks post-conception (WPC), according to the foot-length criterion. Presently, the foot-length criterion is considered the most acceptable parameter used to calculate the gestational age [11-15]. The fetuses were also evaluated regarding crown-rump length and body weight immediately before dissection, and all measurements were taken by the same observer.

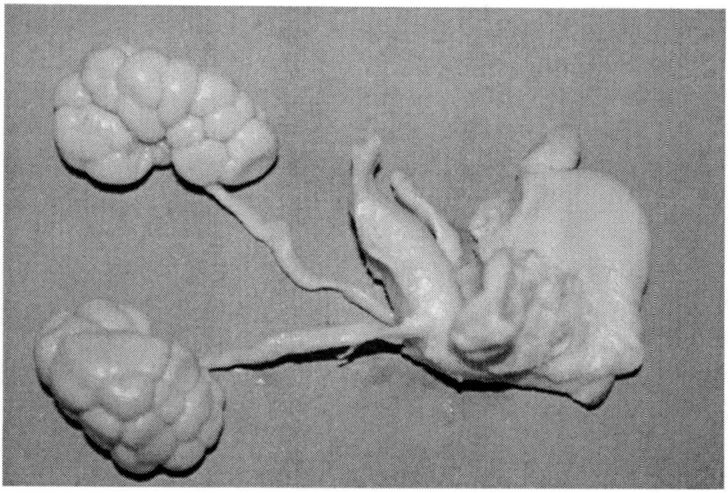

Figure 2. The figure shows the urogenital organs of a male fetus with 20 weeks post-conception of the control group.

After the measurements, the fetuses were carefully dissected with the aid of a stereoscopic lens with 16/25X magnification. The urogenital organs were carefully removed (Figure 2). The ureters, the bladder and the penis were separated from the other structures and fixed in 10% buffered formalin, and routinely processed for paraffin embedding, and 5μm thick sections were obtained at 200μm-intervals. Smooth muscle, connective tissue, elastic system fibers and collagen were studied by histochemical, immunohistochemical and biochemical methods.

Sections were stained with haematoxylin-eosin to assess the integrity of the tissue. We performed the following staining: Masson's trichrome, in order to quantify connective tissue and smooth muscle; Weigert Resorcin Fucsin with previous oxidation in order to observe elastic system fibers; and picro-Sirius Red with polarization for observation of different collagen types. Connective tissue, smooth muscle and elastic system fibers were quantified by a stereological method [16].

Five sections were stained, and five fields of each section were selected. All selected fields were photographed and the images were captured with Olympus BX51 microscopy and Olympus DP70 camera. Images were transferred to Image Pro software. The fibers were quantified using software Image J in order to determine the volumetric density (Vv) of each component.

The immunohistochemical analysis of the collagen type III (mouse monoclonal collagen III ABCAM) and collagen type I (mouse monoclonal collagen I ABCAM) fibers used the avidin biotin (ABC) method with positive and negative controls. The slides were previously treated with poly-L-lysine for better adherence of the sections.

For the biochemical analysis of the collagen, tissue samples were fixed in acetone. The concentration of total collagen in the bladder tissue was determined by a colorimetric hydroxyproline assay. Thus, 5 to 14 mg of dry, defatted bladder tissue was hydrolyzed in 6N HCl for 18 hours at 118°C, as previously described [17]. The assay was then carried out in the neutralized hydrolysates using a chloramine T method [18]. Results were expressed as micrograms of hydroxyproline per milligram of dry, defatted tissue.

Means were statistically compared using the Unpaired T test ($p<0.05$) with Graph Pad Prism software.

Structure of the Urogenital Organs

The age of the normal fetuses studied ranged from 14 to 23 WPC, they weighed between 125 and 950g and had crown-rump length between 13 and 26.5 cm. Anencephalic fetuses ranged in age between 18 to 22 WPC, weighed between 170 and 420g and had crown-rump length between 13 and 16.5 cm.

Ureter

We did not observe elastic system fibers in any fetal ureters and bladder analyzed. There was no difference between gender, anencephalic and control groups, when calculating age, weight, ureteral wall thickness and lumen area. Using simple linear regression we could observe that ureteral lumen area ($R^2 = 0.52$) and wall thickness ($R^2=0.73$) increases gradually with ages advance, independent of the presence of anencephaly. This relation was not observed when using weight as a parameter ($R^2= 0.31$).

The ureteral epithelium was well preserved in the anencephalic and control groups. We did not observe differences in the transitional epithelium in the anencephalic and control groups. The analysis of the ureteral lumen area shows a decrease ($p=0.0047$) in anencephalic ureter ($6365 \pm 1282\mu$) as compared to normal ($20170 \pm 5480\mu$). The ureteral diameter shows a decrease ($p=0.0024$) in anencephalic ureter ($166.7 \pm 10.99\mu$) as compared to normal ($240.0 \pm 26.66\mu$) and ureteral thickness were significantly smaller ($p=0.0144$) in anencephalic fetuses ($30.57 \pm 2.034\ \mu$) as compared to normal (47.49 ±

7.453 μ). Stereology analysis documented a decrease without significance ($p=0.1215$) of SMC in anencephalic ureter ($12.00 \pm 1.628\%$) as compared to normal ($13.51 \pm 0.9231\%$). We can see the structure of the ureter in figure 3.

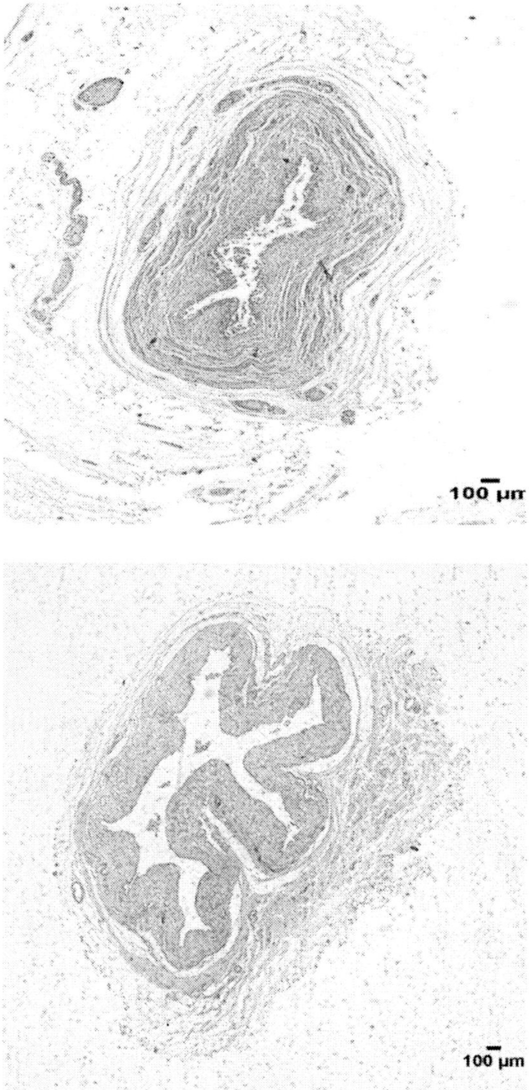

Figure 3. Structure of the ureter. The ureteral diameter and thickness were significantly smaller in anencephalic fetuses as compared to normal. A) Ureter of fetus with anencephaly with 30 WPC. HE x 40. B) Normal fetal ureter with 27 WPC. HE x 40.

Bladder

Stereological analysis documented a statistically significant increase (p<0.0001) of connective tissue in anencephalic bladder. When we compared the gestational age with connective tissue, we did not find any correlation either in normal or in anencephalic fetal bladder. Stereology analysis documented a statistically significant decrease (p<0.0001) of smooth muscle in anencephalic bladder as compared to normal. When we compared the gestational age with smooth muscle, we did not find any correlation either in normal or in anencephalic fetal bladder. In figure 4 we observe photomicrographies when comparing the smooth muscle arrangement in normal and anencephalic fetus bladders.

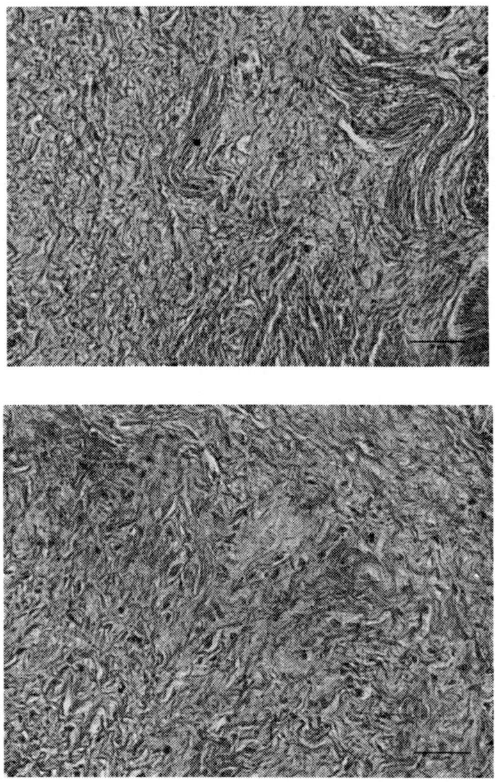

Figure 4. Photomicrographies showing connective tissue and smooth muscle of the fetal bladder. A) Normal fetal bladder with 18 WPC. Masson's trichrome x200. B) Bladder of fetus with anencephaly with 18 WPC. We can observe a reduction of smooth muscle in anencephaly fetal bladder. Masson's trichrome x200.

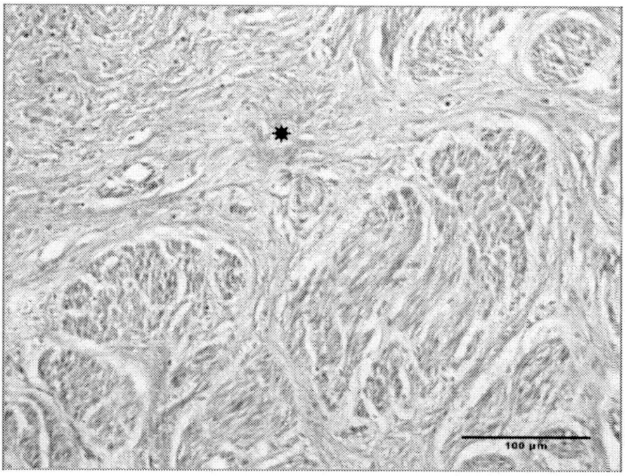

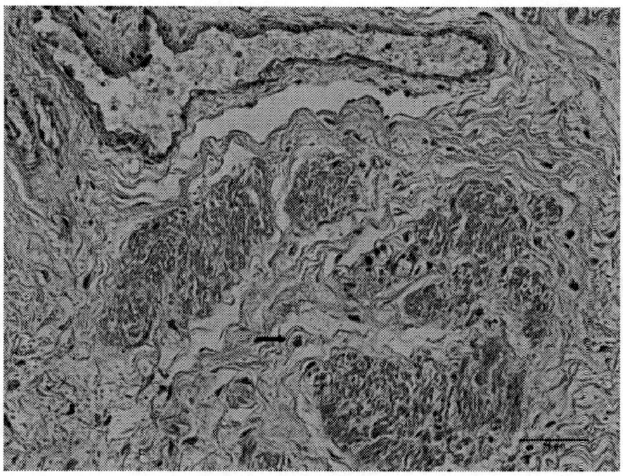

Figure 5. Photomicrographies showing fetal Bladder collagen. A) Immunohistochemistry showing type III collagen (*) in a normal fetal bladder with 17 WPC. Anti-collagen type III antibody X400. B) Immunohistochemistry showing type III collagen (brown) in anencephaly female fetal bladder with in a fetus with 16 WPC. Anti-collagen type III antibody X400.

Biochemical analysis showed a statistically significant increase ($p < 0.02$) in total collagen concentration in the anencephalic group (48117µg/mg; 95%CI: 32.38 ± 42.33) as compared to the normal group (37354µg/mg; 95%CI: 39.71 ± 56.52). In qualitative analysis, type III collagen was observed in both groups, although anencephalic bladder fetuses showed a higher quantity (Figure 5). Regarding type I collagen, a small quantity was observed in both groups [19].

Penis and Urethra

Tunica albuginea and the intracavernosal septum were present in all samples (Figure 6). There was no difference too in structure of the urethra in normal and anencephalic fetuses [20]. We did not observed differences between SMC in corpora cavernosa and spongiosum comparing anencephalic (p=0.54) to normal fetuses (p=0.16). Immunolabeling confirmed previous results showing that the cavernosa trabecula in anencephalic fetuses has a very similar distribution to the normal one (Figure 7).

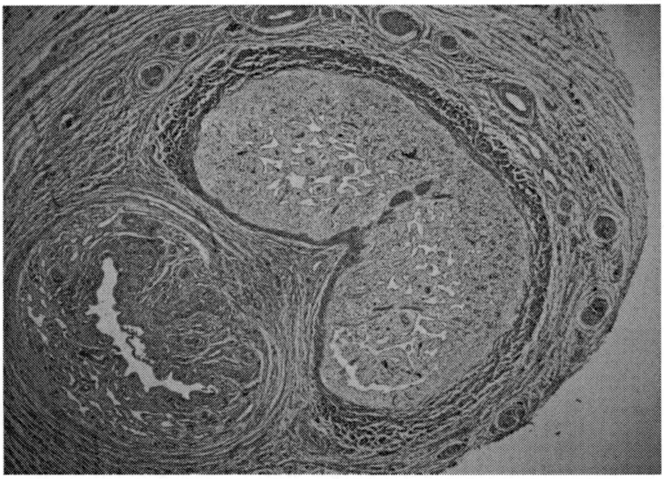

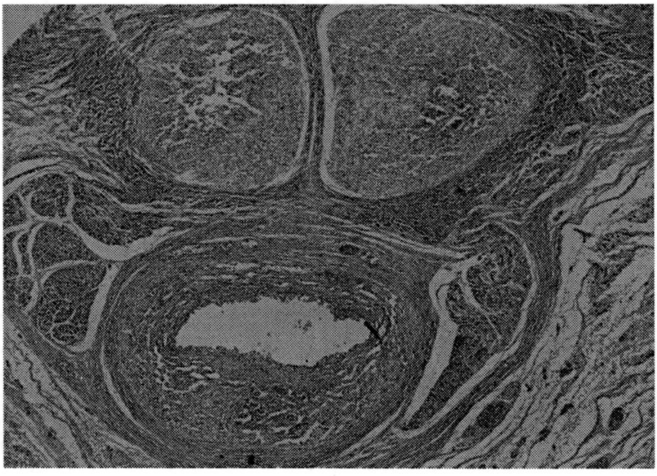

Figure 6. (Continued).

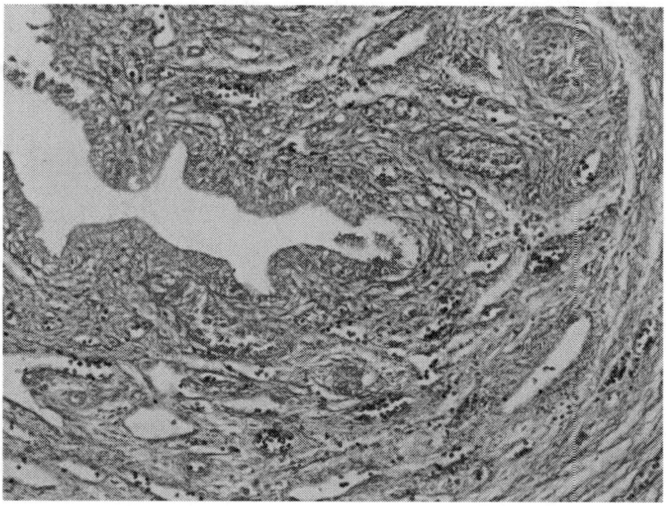

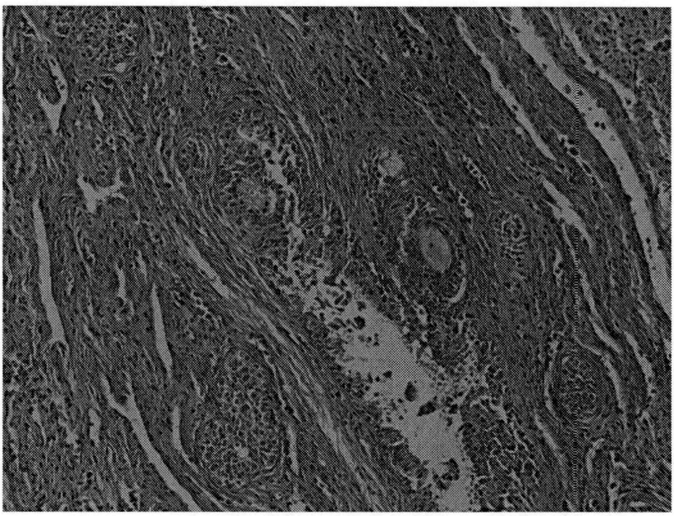

Figure 6. Photomicrographies showing the fetal penis and urethra. A) Penis of a normal fetus with 22 WPC. We can see the corpora cavernosa and the corpora spongiosum. Masson's Trichrome X200. B) Penis of a fetus with anencephaly with 30 WPC. We can see the corpora cavernosa and the tunica albuginea with tissue distribution similar to normal penile fetuses. C) Penis of a normal fetus with 22 WPC showing the corpora spongiosum and urethra. Masson's Trichrome X200. D) Penis of a anencephalic fetus with 33 WPC showing the corpora spongiosum and urethra. Masson's Trichrome X200. There was no difference too in structure of the urethra in normal and anencephalic fetuses.

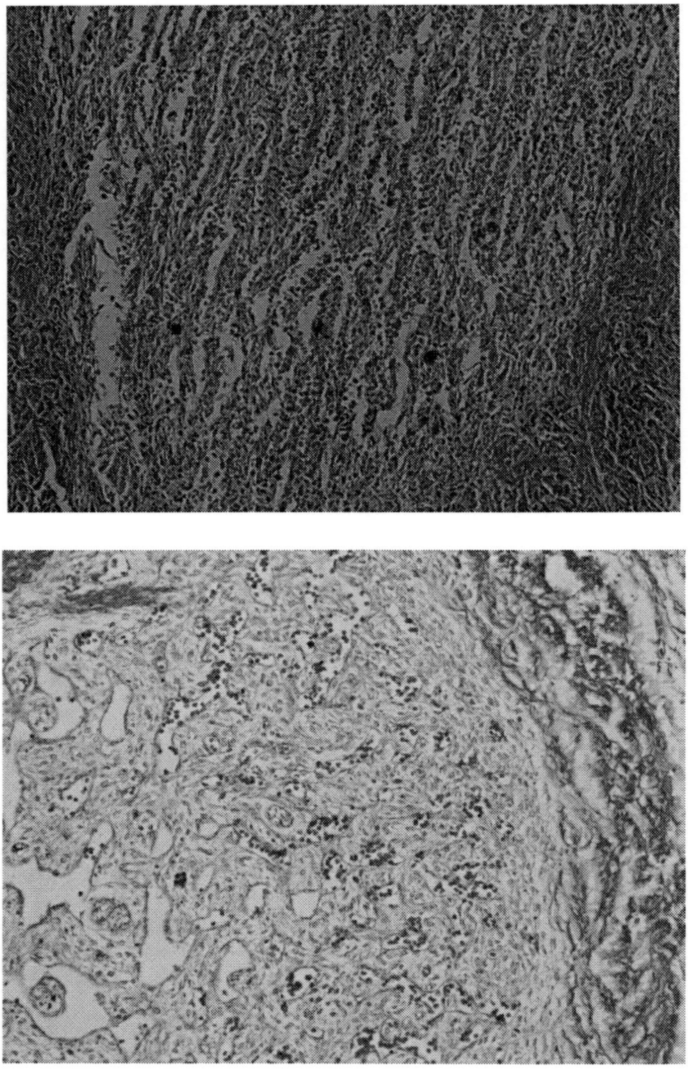

Figure 7. Photomicrographies showing connective tissue and smooth muscle of the fetal penis. A) Penis of a normal fetus with 20 WPC. We can see the corpora cavernosa (Cc) and the tunica albuginea (ta). Masson's Trichrome X200. B) Penis of a fetus with anencephaly with 21 WPC. We can see the corpora cavernosa (Cc) and the tunica albuginea (ta) with tissue distribution similar to normal penile fetuses.

In corpora cavernosa, applying simple linear regression, Cot increases with the advance of age and SMC decreases, regardless anencephaly (p<0.01

for both samples). These findings did not occurred in the spongiosum, once there was no difference in elements SMC or Cot *versus* age.

Histochemistry with Weigert's has shown unclear results and to confirm the presence of elastic system elements, immunolabeling with elastin has been performed, showing its positivity only in fetuses from 20th WPC. There was no difference in elastic fibers distribution in normal and anencephalic fetuses.

Comments

In our samples we do not observed significant differences in SMC of the ureter when comparing anencephalic and normal fetuses. A slight but not significant increase in SMC was observed in normal fetuses, but we observed significant decrease of lumen area, thickness and diameter in ureter of anencephalic fetuses. In a recent study we observed gross histological alterations in bladder of anencephalic fetuses [19]. The lesion in the nervous system with consequent alteration in nerve regulation could be a plausible hypothesis to explain structural changes in bladder and ureter in patients with neural tube defects.

Anencephalic fetuses have cerebral exposition, usually with spinal cord preservation. Bladder nerves in anencephalic fetuses could be modified due to cerebral lesions with consequent brain control damage in bladder nerves. This could lead to structural alterations in anencephalic bladder fetuses and in the ureters too.

Collagen and elastin are important components of the ureter wall, which affect ureter contraction. Collagen provides tensile strength although an over accumulation may inhibit ureteral contractility and the conduction of electrical impulses through the wall. Elastin provides tissue elasticity and could help compliance [21].

In our study, we observed a significant increase of collagen in anencephalic ureter fetuses with type III collagen predominance. In previous study of anencephalic bladder we observed the same distribution of collagen with type III collagen predominance. Collagen increases in bladder chronic obstruction in human adults. The abnormally large amount of collagen found in obstructed bladders is also believed to decrease muscle contractility, in addition to affecting bladder compliance and subsequent high intravesical pressure, which, result in changes in ureteral contraction and leads a histological changes in ureteral wall [21]. Previous studies in rabbits, shows that ureteral obstruction causes progressive thickening of the lamina muscular with a great increase in collagen [22].

In the bladder of anencephalic fetuses we observed a greater amount of connective tissue than in the normal fetuses' bladders. It could suggest an alteration in the complete development of the smooth muscle layer.

Elastic system fiber alterations are involved in fibrotic tissue formation; however, in our samples, we did not observe the presence of elastic fibers in the bladder. This may indicate that this extracellular matrix component appears only in the third gestational trimester in fetal bladder. Previous studies showed the elastic system fibers in other human fetal genitourinary organs [23].

Bastos [23] observed scarce and fine elastic system fibers in the homogeneous and intense cellular tissue of the corpus spongiosum, in a fetus with 15 weeks of gestation; in a fetus with 36 weeks of gestation, there was evidence of the trabeculae of the corpus spongiosum delimitating large vascular spaces. The elastic system fibers are plentiful and organized in older fetuses [23]. This study indicates that the elastic system fibers in the genitourinary fetal system are more evident and developed in the third gestational trimester. Our sample was composed of fetuses in the second gestational trimester, probably the period where elastic system fibers are still being formed in fetal bladder.

Structural bladder alterations in anencephalic fetuses were significant in our study. The lesion in the nervous system with consequent alteration in bladder nerve regulation could be a plausible hypothesis to explain these structural changes. Anencephalic fetuses have cerebral exposition, usually with spinal cord preservation. Bladder nerves in anencephalic fetuses could be modified due to cerebral lesions with consequent brain control damage in bladder nerves. This could lead to structural alterations in anencephalic bladder fetuses.

Interestingly, there is an increase of its AD with the advance of age in normal and anencephalic fetuses and no difference was found when comparing both groups. These results suggest that the turnover of Cot is a physiologic aspect of corpora cavernosa in human development and cannot be compared to animal samples.

SMC decreased in older fetuses (normal and anencephalic). Once again no difference was observed when comparing groups. A slight but not significant increase was observed in normal fetuses. Trough myosine immunolabeling we have observed particularities in SMC distribution in Corpora Cavernosa trabeculae. SMC are more superficial in this tissue than in spongiosum. The decreases of SMC and increases of Cot suggest that there is a gradual substitution of SMC by novel elements as age advances.

We did not observe any correlation between tissue's distribution (Cot, SMC and EF) and age in corpus spongiosum specifically. This data indeed gives us a positive aspect of the urinary tract development. Probably urethral development is maintained and unchanged in anencephalic fetuses when compared to normal. This data becomes useful when comparing to previous studies focused in urethral reconstruction in adult humans [24]. In the future, tissue from urethras from anencephalic fetuses could be used as engineering models in pediatric population.

Conclusion

The structural alterations of the bladder found in this study may suggest the existence of functional alterations in the bladder of anencephalic human fetuses. There was no difference in the structure of the genitalia of anencephalic fetuses compared to normal ones. Elastin is documented from 20th wpc wich suggests the maintenance of erectile function. Histochemistry and immunolabeling suggests that penile shaft development is maintained and unaltered in NTD fetuses. Although anencephaly is a disease that is incompatible with babies' long spam survivors, this present study shows that ureteral development is unchanged. In our opinion, its anatomy did not suffer any alteration in consequence of this disease and, in the future, this tissue could be used as a substitute for native tissue.

References

[1] Blatter BM., van der Star M, Roeleveld N. Review of neural tube defects: risk factors in parental occupation and the environment. *Environmental Health Perspect*, 1994; 102, 140-145.

[2] Cook RJ., Erdman JN., Hevia M, Dickens BM. Prenatal management of anencephaly. *Int. J. Gynec. and Obst.* 2008;.102, 304-308.

[3] Müller, L., K. Abrahamsson, U. Sillén, B. Jacobsson, O. M. Hellström. Ultrasound assessment of detrusor thickness in children and young adults with myelomeningocele. *J. Urol.* 2006; 175: 704-708.

[4] Calzolari F, Gambi B, Ganari G, Tamisari L. Anencephaly: MRI findings and pathogenetic theories. *Pediatr. Radiol.*, 2004; 34, 1012-1016.

[5] Laberge, JM. Transplanting organs from anencephalic infants. *CMAJ* 1987; 137:437-438.

[6] Salaman JR. Anencephalic organ donors. *BMJ* 1989; 298, 622-623.

[7] Byrne P. Use of anencephalic newborns as organ donors. Paediatr. Child. *Health* 1995; 10, 335-337.

[8] Milliez, J. Anencephaly and organ transplantation. *International Journal of Gynecology and Obstetrics* 2008; 102: 99.

[9] Koifman L, Vides AJ, Koifman N, Carvalho JP, Ornellas AA; Epidemiological aspects of penile cancer in Rio de Janeiro: evaluation of 230 cases. *Int. Braz. J. Urol.* 2011; 37, 231-243.

[10] Badawy H, Soliman A, Ouf A, Hammad A, Orabi S, Hanno A; Progressive hair coil penile tourniquet syndrome: multicenter experience with 25 cases. *J. Pediatr Surg.* 2010; 45:1514-1518.

[11] Hern WM. Correlation of fetal age and measurements between 10 and 26 weeks of gestation. *Obst. Gynecol.*; 1984. 63. 26-32.

[12] Mercer, BM., Sklar S, Shariatmadar A, Gillieson MS, D'Alton MK. Fetal foot length as a predictor of gestational age. *Am. J. Obst. Gynecol.* 1987; 156: 350-356.

[13] Platt, L.D., Medearis, A.L., DeVore, G.R., Horenstein, J.M., Carlson, D.E., Brar, H.S. Fetal foot length: Relationship to menstrual age and fetal measurements in the second trimester. *Obstet. Gynecol.* 1988; 71: 526-531.

[14] Favorito, L.A., Cardinot, T.M., Morais, A.R.M., Sampaio, F.J.B. Urogenital anomalies in human male fetuses. *Early Human Development* 2004; 79: 41-47.

[15] Costa W. S., Sampaio F. J. B., Favorito L. A. and Cardoso L. E. Testicular migration: remodeling of connective tissue and muscle cells in human gubernaculum testis. *J. Urol.* 2002; 167, 2171- 2176.

[16] Cavalcanti, A.G., Costa, W.S., Baskim, L.S., McAninch, J.A., Sampaio, F.J. A morphometric analysis of bulbar urethral strictures. *BJU Int.* 2007; 100: 397-402.

[17] Cabral C. A. P., Sampaio F. J. B. and Cardoso L. E. M. Analysis of the modifications in the composition of bladder glycosaminoglycan and collagen as a consequence of changes in sex hormones associated with puberty or oophorectomy in female rats. *J. Urol.* 2003; 170, 2512- 2516.

[18] Bergman I. and Loxley R. Two improved and simplified methods for the spectrophotometric determination of hydroxyproline. *Anal. Biochem* 1963; 35, 1961- 1965.

[19] Pazos HM, Lobo ML, Costa WS, Sampaio FJ, Cardoso LE, Favorito LA: Do neural tube defects lead to structural alterations in the human bladder? *Histol. Histopathol.* 2011;26:581-8.

[20] Carvalho, J.P.M., Costa, W.S., Sampaio, F.J., Favorito, L.A. Anencephaly does not cause structural alterations in the fetal penis *J. Sex Med.* 2012; 9: 735-742.

[21] Kim, K. M., B. A. Kogan, C. A. Massad, Y. C. Huang. Collagen and elastin in the normal fetal bladder. *J. Urol.* 1991; 146: 524-527.

[22] Cheng EY, Maizels M, Chou P, Hartanto V, Shapiro E. Response of the newborn ureteropelvic junction complex to induced and later reversed partial ureteral obstruction in the rabbit model. *J. Urol.* 1993; 150: 782-789.

[23] Bastos A. L., Silva E. A., Costa W. S. and Sampaio F. J. B. The concentration of elastic fibers in the male urethra during human fetal development. *BJU. Int.* 2004; 94, 620-623.

[24] Raya-Rivera A, Esquiliano DR, Yoo JJ, Lopez-Bayghen E, Soker S, Atala A. Tissue-engineered autologous urethras for patients who need reconstruction: an observational study. http://www.ncbi.nlm.nih.gov/ pubmed/21388673 2011; 377: 1175-1182.

Index

D

E

F

M

N

O

P

Q

R

S

T

U

V

W

Y

Z